QUALITY LABS FOR SMALL BREWERS

BUILDING A FOUNDATION FOR GREAT BEER

MERRITT WALDRON

Brewers Publications®
A Division of the Brewers Association
PO Box 1679, Boulder, Colorado 80306-1679
BrewersAssociation.org
BrewersPublications.com

Proudly Printed in the United States of America.
10 9 8 7 6 5 4 3 2 1
ISBN-13: 978-1-938469-63-3
ISBN-10: 1-938469-63-1
Ebook ISBN-13: 978-1-938469-64-0

Library of Congress Control Number: 2020940188

Library of Congress Cataloging-in-Publication Data is on file with the Library of Congress.
The following is for reference only:

Name: Waldron, Merritt
Title: Quality Labs for Small Brewers: Building a Foundation for Great Beer
Identifiers: ISBN 978-1-938469-63-3; EISBN: 978-1-938469-64-0
Subjects: quality control; best practices; labratory testing; craft beer; protocols; quality assurance

Publisher: Kristi Switzer
Copyediting: Iain Cox
Technical Editor: Rob Christiansen and Jason Perkins
Indexing: Doug Easton
Art Direction: Jason Smith
Interior Design: Danny Harms
Production: Justin Petersen and Kate Sanford

To Adrienne and the rest of my family for your abiding love and support. Nathan and Heather for taking a chance on me and guiding the ship.

TABLE OF CONTENTS

FOREWORD

For the past two years I have had the pleasure of teaching the Brewers Association's Basics of Beer Quality course, a one-day laboratory boot camp filled with lectures and hands-on activities. The goal is to educate brewers from average-size breweries in the USA, ones making approximately 600 barrels of beer annually, and give those brewers the information they need to start a quality program to support production or build upon their existing foundation. I like to call it the "Traveling Science Road Show." Anyone who has attended this workshop can attest to the fast-paced nature of the course. There is only so much information my colleagues and I can successfully pass on in the eight limited hours of time we have together, and, unfortunately, there are always topics and stories we do not have time to discuss. With that in mind, I was elated to read Merritt Waldron's *Quality Labs for Small Brewers: Building a Foundation for Great Beer*.

Quality Labs for Small Brewers stands apart from the crowd, being both an informative introduction for the novice brewer and a valuable reference guide for the brewing quality professional. Much of the content it provides dovetails nicely with the Basics of Beer Quality Workshop and other resources in our industry. I love the depth and attention this book is able to apply to quality concepts that we can only otherwise introduce in the confines of a one-day course. I truly believe *Quality Labs for Small Brewers* is essential reading for any brewery employee who wants to dive deeper into quality-related content. This book tackles many challenging topics, such as writing a quality manual, troubleshooting instrumentation, creating control charts, and more, all in a digestible, interesting format.

I have a secret to share: I entered the beer industry as a minority woman in the sciences who knew nothing about beer. I knew so little about beer, in fact, that I did not even drink beer. In my early days at the brewery, I set out to apply the skills I had from my chemistry degree to design and run experiments, collect data, and tell a story with that information to the brewery. Why am I recounting these early days? Because quality control can feel daunting. The truth is, when I started, I felt like an imposter—one who did not know the jargon or understand the details of the beer industry.

Quality is often considered an intimidating concept for brewers who do not come from a traditional science background. Many of the brewers I have met while teaching arrive to class thinking beer quality work is something only scientists tackle. Permit me to disagree. My work has allowed me to meet so many brewery industry folks who are scrappy, smart, and hungry for information. Through my work as an instructor, I have seen the clouds lift from people's eyes on topics they previously felt were beyond their comprehension. I am here to say that I believe in every brewer's capability to build and execute a quality program within their walls. Merritt has created an indispensable resource for a seemingly impossible journey, directed at this very audience. *Quality Labs for Small Brewers* provides a roadmap for taking yourself and your brewery to the next level. Reading this book is an important first step toward becoming acclimated to a new world rife with a unique and challenging vernacular. Quality programs are rarely built in a day, especially in larger breweries. When we made the commitment at Bell's Brewery to put our quality plan into writing, we knew it would take years and become a living document. The first draft was far from perfect, but it gets a little better every time we review it.

Walking the quality line is not always an easy path to travel. Quality professionals are often the bearers of bad news; for instance, we have to make recommendations on whether or not to destroy beer. It is easy to feel the weight of this decision when considering the raw material costs, time, and labor that go into brewing a beer. Furthermore, we do not always get to have the final say on what happens to a quality-compromised beer. There have been times in my career where a batch has been released to package against my recommendation. I know that the powers that be will not always agree with my conservative approach, and those are often my hardest days. While the experience can be discouraging, I make the most of these moments by continuing to collect valuable data to help address the underlying problems should they occur in the future.

Merritt's book contains detailed explanations critical to laying the foundation of your quality program. I know that having this book in my early days of trouble-shooting would have been extraordinarily helpful. That said, one facet of the brewing industry I truly love is that you are not alone. Once indoctrinated into the secret society of brewing professionals, the available resources are vast. The written content, the audio content, and the people who want to discuss and solve problems are endless. The greater brewing community's willingness to share is something that makes this industry so unique and special. Speaking to quality professionals in other industries has taught me that we should not take this community for granted. This communal effort toward improvement is palpable, and nothing could be more in this spirit than writing a book such as this one.

I like to end my classes by saying beer is magic. Your brewery can have all the bells and whistles and make terrible beer, or your brewery can have nothing and make award-winning beer. Without a doubt, committing to and investing the time in a quality program brings us all closer to the perfect pint the first time around.

Cheers!

Lauren Torres
Laboratory Manager, Bell's Brewery, Inc.

ACKNOWLEDGMENTS

Many of my peers graciously took time to help me though this project. Without them none of this would have been possible and I thank them: Alex, Missy, Brady, and the rest of the Rising Tide Brewing family; Jason Perkins, Zach Bodah, Mike Billon, and the Allagash Brewing Company lab team; and Lindsay Barr, Rob Christiansen, Chuck Skypeck, Jamie Floyd, and the Brewers Association quality subcommittee. My thanks also go to Campbell Morrissy (Mother Road Brewing Company), Ben Parsons (Baerlic Brewing Company), Julie Smith (Lawsons Finest Liquids), Andy Tveekrem (Market Garden Brewery), Jared Kiraly and Liz Kiraly (Bone Up Brewing Company), Christie Mahaffey (Foundation Brewing Company), Amy Todd (Zero Gravity Craft Brewery), Jason Bolton, PhD (University of Maine Cooperative Extension), and all of the brilliant brewers and scientists that came before me.

INTRODUCTION

What a wonderful time to be to be in the craft brewing industry. Growth is continuing, consumers' tastes are varied and discerning, and the professional community is better than any other industry on earth. There is never a dull moment. The industry is continuously pushing the boundaries of craft beer, whether it is making beers bigger and better, reviving lost styles, or inventing new ones. As the craft brewing industry strengthens and matures, so must the small brewer. We must strive to be continuously improving. Not only for ourselves but also because what we do reflects on the industry as a whole.

You have already come so far. You learned how to make great beer and opened the doors to your brewery or maybe finally got your foot in the door of your favorite brewery. But the journey to continuous improvement is a long one. Your next challenge lies within these pages.

As the craft brewing industry grows so do the stakes. With so many moving pieces and so much industry growth, having a Quality Assurance and Quality Control (QA/QC) System is no longer an option but a necessity for brewers of all sizes. An effective QA/QC System can set you apart from the competition, helping you troubleshoot issues and improve batch consistency, thus building a loyal consumer base and generating revenue.

Most importantly, your QA/QC System acts as an insurance policy. It can prevent bad, even potentially dangerous, beer from reaching the consumer. Recalls can destroy your reputation with consumers and retail partners, halt sales, and bankrupt your brewery.

Well, that was scary… but I am here to tell you it does not have to be. It is with you—the small, independent brewer—in mind that I have written these pages. I will equip you with the tools and concepts you need to grow up to 10,000 barrels per year and beyond while having some fun along the way. If growth is not your thing, the concepts that follow will refine your beer, empower you and your staff, and allow you to understand your process at a deeper level.

ABOUT THE BOOK

This is not a book on how to brew—you probably already know how to do that. This is not a book about fermentation, because that has already been written. This book is intended to help small craft brewers put an effective QA/QC system in place from the ground up. You will not just be measuring stuff and hoping the brew goes well. You will implement policy and systems that enhance the value of the measurements you are making. In this book, we will look at what you are measuring and how it works, and how to gather your measurement data and put it to work so that you get better and more consistent beer as a result.

What I Assume about the Reader

When writing this book I tried to have a target reader in mind. First, I reflected on when I started my first QA/QC job at a craft brewery and wrote about what I wish I had known back then. I also spoke to some colleagues in the industry and learned that the average production output for craft brewers is around 600 barrels. With both of these things in mind, I have assumed that you:

- have a good understanding of brewing processes and principles;
- are looking to improve the consistency or overall quality of your beer;
- want to start QA/QC or is supervising someone in charge of it;
- have a basic understanding of algebra;
- have a basic understanding of spreadsheet formulas and charts;
- are working with a small-to-modest budget; and, most importantly,
- are very curious.

There is a lot to learn in the pages that follow. To fit it all in, a lot of the concepts are presented at the surface level as it is relevant to your brewing process. In most instances there is much more to learn, and I encourage you to keep digging once you finish reading this book.

THE STRUCTURE OF THE BOOK

This book is set up in four parts: Building a Solid Foundation, Building Your Quality Program, Measurements, and Getting to Work. Part 1, "Building a Solid Foundation," outlines some prerequisite systems that your brewery is required to implement under the Food Safety Modernization Act of 2011. This is a perfect start for your quality system as these prerequisite programs lay the foundation for producing a wholesome and safe food product (i.e., beer).

Part 2, "Building Your Quality Program," builds on the concepts introduced in part 1, walking you through the setting up of your quality plan and writing a quality manual, which you will use as a handbook for implementing policy, training staff, and the practical application of quality-focused brewing.

Part 3, "Measurements," goes hand in hand with part 2. The measurements you take are informed by your quality system and those benchmarks and influencers that come to inform the inevitable growth and evolution of your quality plan. In an effort to pull back the veil and understand what the numbers mean, the principles of each measurement are discussed. We will look at the measuring principles of available instruments, point out useful methods, and suggest quality control points (QCPs) to include in your quality plan.

Part 4, "Getting to Work," will show you how to put your measurement data to work. You will organize, analyze, build confidence in, and apply data to your brewing processes to effect change as needed.

I suggest reading this book through the first time to gain an overall view of a craft brewery quality system and how the threads weave together. However, I do not expect you to apply the information and ideas in such a linear fashion. While our goals may be the same there is more than one road to take, so start with the low-hanging fruit and start taking immediate steps toward implementing or improving your brewery's quality system. Without further ado, let us get down to the task at hand.

PART I
BUILDING A SOLID FOUNDATION

When I tell folks that my job is Director of Quality Control and Assurance at a brewery their immediate response is, "You must have the best job in the world, you get to drink beer all day." The first part of that statement is 100% correct, it is the best job in the world. The last part contains some truth, but most of my duties do not involve imbibing copious amounts of beer. Since quality control and assurance is an abstract topic to many, let us define what it means in the brewery setting.

QUALITY CONTROL

To put it in simple terms, quality control is a measurement process. You have most likely been taking measurements since your homebrewing days. This process can be anything: weighing pounds of grain, measuring the volume of water in your mash, taking the temperature of wort coming out of your heat exchanger, monitoring the daily apparent extract of fermenting beer, and measuring volumes of CO_2 in final package. Countless direct and indirect measurements can be taken throughout your brewing process. While there are measurements that are "non-negotiable" for production breweries, and I will get to those later in this book, it is up to you to determine what measurements are important to your brewery process and size.

QUALITY ASSURANCE

Often overlooked by many start-up breweries and homebrewers alike, quality assurance is making sure that your measurements are accurate. Accurate measurements are the most important part of a QA/QC system and the validation and calibration of your measurement tools and equipment is critical. Not every hydrometer or thermometer is created equal and results can vary between instruments. This is a problem if you are using more than one of those instruments in your brewery or if an instrument breaks (because it will break).

If quality assurance seems intimidating to you, remember to start simple. For example, check your reading against a known standard. In the case of a hydrometer, use distilled water at 68°F. What do you get, is it 1.000 specific gravity (SG)? Good. Next, try a simple solution of 1 gallon distilled water and 1 pound dextrose. Does it read 1.037 SG? Excellent. Now cut that solution in half with distilled water or try densities of simple solutions—do the readings make sense? If so, you are well on your way to validating that instrument. Not a big deal, right? In order to validate your instrumentation, you should understand what you are measuring and the principles under which your instruments work. We will cover a lot of this later. Stay tuned.

QUALITY SYSTEM

The quality system is a set of rules and procedures that ensure a quality product is being produced. This is where it all starts to come together. Within your quality system you determine when and where your process measurements are taken, which are your quality control points (QCPs). Once QCPs are identified and measured successfully, set up specifications and standards for those measurements, including how often you calibrate your instrumentation. To accomplish this, you will write a quality manual to support your quality system, which will include the points mentioned above and will help determine what actions take place when there is a result out of specification. Out-of-specification results are usually due to human error or mechanical failure. Mistakes happen, it is human nature. None of us is infallible, but putting in place a robust quality system can significantly reduce the frequency and magnitude of these mistakes.

On the craft brewing level, QA/QC is not about having all the big, fancy scientific equipment. It is about having and implementing a few simple tools and policies effectively. What those tools are will depend on what you are trying to accomplish. Even more important, QA/QC is about having thoughtful, well-implemented policies and organization. When you have a strong QA/QC foundation and use some simple tools effectively you can accomplish great things. Once your system is in place and your operation expands, you can start adding scientific equipment and employing staff to run it. But you would be surprised how much you can get done with a good hydrometer, thermometer, and pH meter, some used equipment, a bit of duct tape, and a few pints of beer.

1

FOOD SAFETY MODERNIZATION ACT

Why is this book on small brewery quality control starting with an act of Congress? To put it simply, the passing of the Food Safety Modernization Act (FSMA) has shifted the way the Food and Drug Administration (FDA) handles food manufacturing. At the same time, the FSMA specifically defines alcoholic beverages as food, so, no matter what size your brewery is, you must comply with parts of the FSMA. When the act was passed in 2011, there was some confusion and debate among independent craft brewers over who needed to comply and with what parts of the act. However, even a "very small business" (i.e., a business with average annual sales of less than $1 million) had to be FSMA compliant as of September 17, 2018.

That means sometime soon your brewery may get a visit from an FDA or FDA-sanctioned inspector assessing your FSMA compliance. While most brewers I have spoken to have heard of FSMA, not many know what it means to their operation or how to comply. By discussing the impetus behind the FSMA and what it covers, you and your brewery will be in a good position to formulate policy for compliance. At the end of the day, food safety is everyone's responsibility. I would hate to see the image of craft brewing tarnished by a few bad incidents.

INTRODUCTION TO THE FSMA

According to recent data from the Centers for Disease Control and Prevention, each year about 48 million people in the US get sick, 128,000 are hospitalized, and 3,000 die from foodborne diseases. That is a staggering one in six people affected each year. Recognizing that preventable foodborne illness is both a significant public health problem and a threat to the economic well-being of the food industry, the 111th Congress amended the Federal Food, Drug, and Cosmetic Act with respect to the safety of the food supply with the FSMA. The main goal of this reform was to shift the FDA's focus from responding to foodborne illness to prevention of foodborne illness.

Recognizing that ensuring the safety of the food supply is a shared responsibility among the actors at various points in the global supply chain for both human and animal food, the FDA has finalized seven major rules to implement the FSMA. These rules are designed to make clear specific actions that must be taken at each of these points to prevent contamination.[1] These rules can be found in the *Code of Federal Regulations*, 21 C.F.R. 117. They are outlined as follows:

1 "FSMA Final Rule for Preventive Controls for Human Food - Current Good Manufacturing Practice, Hazard Analysis, and Risk-Based Preventive Controls for Human Food," U.S. Food and Drug Administration (website), accessed September 20, 2018, https://www.fda.gov/Food/GuidanceRegulation/FSMA/ucm334115.htm.

21 C.F.R. 117: Current Good Manufacturing Practice, Hazard Analysis, and Risk-Based Preventive Controls for Human Food	
Subpart A	General Provisions
Subpart B	Current Good Manufacturing Practice
Subpart C	Hazard Analysis and Risk-Based Preventive Controls
Subpart D	Modified Requirements
Subpart E	Withdrawal of a Qualified Facility Exemption
Subpart F	Requirements Applying to Records That Must Be Established and Maintained
Subpart G	Supply-Chain Program

These seven rules initiate new responsibilities for food companies, new controls over imported food, new powers for the FDA, and new fees on food companies and importers. And, yes, alcoholic beverages are considered a food under 21 U.S.C. 321(f). It is important to recognize that the FSMA specifically defines alcoholic beverages as food, because it brings breweries under FDA regulation. Craft brewing is an exciting industry where creativity abounds, so it is easy to forget that beer is a food product. But food safety should be of primary concern for every staff member in your brewery. The FSMA also outlines several prerequisite programs, Good Manufacturing Practice (GMP) and Food Safety. These programs are the foundation of your Quality Assurance and Quality Control (QA/QC) program.

Breweries are also under the regulatory authority of the Alcohol and Tobacco Tax and Trade Bureau (TTB). If the TTB deems your beer adulterated (a term that is defined below) it can alert the FDA, who can seize your products to stop them from entering commerce. I mention this because it is ultimately the responsibility of management to ensure beer leaves a facility unadulterated. A brewer should be aware of all regulatory agencies involved. In a small craft brewery the responsibility of maintaining compliance with all of these agencies may fall on one person—if that person is you, understanding what requirements the FSMA places on your brewery is imperative.

Regardless of your position at the brewery, compliance with FSMA is a QA/QC manager's best friend because it helps your brewery ensure quality. When planning and implementing a QA/QC plan, the FSMA outlines your most important prerequisite programs. You can use these FSMA subparts as outlines for writing your GMP and food safety programs. I will walk you through writing these programs in the chapters that follow. These programs are the base of your QA/QC program and will set the tone of your "quality first" policy. Just remember that for total company buy-in this must come from the top down. If you are having trouble getting buy-in, you can use the FSMA and its implications and build a case to initiate change from ownership/management.

Food Safety Terms

Before we dig further into the FSMA there are a few terms we should familiarize ourselves with:

Adulteration: To produce, pack, or hold food under insanitary conditions whereby food contains any poisonous or deleterious substance that may render food injurious to health; or that the food contains any unapproved food additive.

Contamination: Adulteration of any product or ingredient with pathogens, chemicals, allergens, radiation, or foreign objects that makes them no longer wholesome and safe.

By not attempting to comply with the FSMA your beer can be considered adulterated. The FDA can initiate seizure of adulterated foods (your beer) to stop them from entering commerce. The first step you can take to avoid this fate is to initiate policies and systems (GMP, food safety and quality) and document your attempts to comply. This basic outline is what we will cover in later chapters.

Lucky for brewers, beer is considered a low-risk food. Due to inherent processing steps (e.g., the boil) and properties (pH under 4.5, an alcohol and CO_2-rich environment), beer is unlikely to support pathogenic microorganisms that can harm humans. This does not mean there is no risk, but it does exempt brewers from some of the seven rules of the FSMA.

WHAT THE FSMA MEANS FOR CRAFT BREWERS

There are a few general points worth highlighting before we delve into specific parts of the FSMA that breweries need to take note of.

- All facilities that manufacture, process, pack, hold, or transport FDA-regulated food products designed for consumption by humans must comply with the FSMA. This includes breweries, wineries, cider

Portions of FSMA that apply to breweries	Portions of FSMA that do NOT apply to breweries*
FDA Facility Registration (21 CFR 1.225)	21 CFR 117 Subpart C – Hazard Analysis and Risk-based Preventive Controls
Compliance with Definitions (21 CFR 117 Subpart A, includes staff competencies and training)	21 CFR 117 Subpart G – Supply Chain Management
Compliance with Good Manufacturing Practices (21 CFR 117 Subpart B)	21 CFR Parts 11 and 121 Protecting Food Against Intentional Adulteration
Compliance with Record keeping requirements (21 CFR 117 Subpart F)	21 CFR Parts 1, 11, and 111 Foreign Supplier Verification Programs

* As per 21 CFR 117.5(i)(1).

Figure 1.1. A glance at important portions of the FSMA that do and do not apply to craft breweries. From "Food Safety Modernization Act FAQs for Brewers," Brewers Association (website), accessed September 15, 2018, https://www.brewersassociation.org/educational-publications/food-safety-modernization-act-faqs-for-brewers/ (subscription required).

producers, and distilleries. **Regardless of your brewery's size, you must comply with the FSMA.**

- Because beer is considered a low-risk food, there are a few exemptions for breweries from the FSMA according to 21 C.F.R. 117.5 (i)(1):
  - Breweries do not have to comply with subpart C, "Hazard Analysis and Risk-Based Preventive Controls," abbreviated to HARPC.
  - Breweries do not have to comply with subpart G, "Supply-Chain Program," assuming both conditions of 21 C.F.R. 117.5 (i)(1) are met.
- If your brewery happens to produce a non-alcoholic beverage you are still exempt from subparts C and G if production is less than 5% of gross revenue, according to 21 C.F.R. 117.5 (i)(2)(ii).
- If certain conditions are met, brewers are also exempt from preparing and implementing a Food Defense Plan. Refer to the "FSMA Final Rule" FDA guidance to see if this applies to you.[2]
- Simplified spent grain handling requirements apply *if* a GMP program is in place.

Now that we have a general idea of what the FSMA is and why it is important, let us look closer at the subparts that directly affect breweries (fig. 1.1). I will not print the actual CFR here, but I will lay out the important points for brewers that are using traditional methods and ingredients. If you think your process may fall outside of this, it is up to you to do your due diligence in determining exactly what applies to your operation.

Subpart A – General Provisions

The meat of subpart A addresses the qualifications of individuals who manufacture, process, pack, or hold food (21 C.F.R. 117.4). This explicitly defines that each individual engaged in manufacturing, processing, packing, or holding food (including temporary and seasonal personnel) must receive training in the principles of food hygiene and food safety. Each individual must have the training or experience necessary (or

FOOD HYGIENE VERSUS FOOD SAFETY

Food hygiene supports your food safety program. **Food hygiene** is the application of known principles and practices that prevent foodborne illness; it can also be referred to as safe food handling. The FDA outlines four general steps for food hygiene:

1. Clean—wash hands and surfaces often
2. Separate—do not cross-contaminate
3. Cook—cook to the right temperature
4. Chill—refrigerate foods promptly

The goal of these food handling steps is to produce safe food. You should apply these principles in various ways through your food safety program.

The principle of **food safety** is to identify control points where food hygiene and other control methods are applied. A food safety program is unique to each facility and will be covered in depth a little later in this text.

[2] "FSMA Final Rule for Preventive Controls for Human Food," FDA website (see n. 1 above); "FSMA Final Rule for Mitigation Strategies to Protect Food Against Intentional Adulteration," U.S. Food and Drug Administration (website), accessed August 20, 2018, https://www.fda.gov/food/food-safety-modernization-act-fsma/fsma-final-rule-mitigation-strategies-protect-food-against-intentional-adulteration.

combination thereof) to produce clean and safe food as appropriate to the individual's assigned duties.

Subpart A also states that the facility's owner, operator, or appointed manager is responsible for ensuring each individual is qualified to perform their assigned duties. In order to do so, supervisory personnel should be equipped with proper education or training. Supervisory personnel are responsible for keeping records that document training, ensuring compliance by individuals, and supervising the production of clean and safe food.

As you can see, subpart A revolves around the training of your brewery's staff. To comply, the training must be documented. This can be as easy as keeping a record of each task and having the employee sign their name and date when they complete their training. For more involved safety training sessions, such as for gas or chemical safety, you can reach out to your suppliers, who will come out to perform staff training and give a quick test at the end to ensure understanding by the individual.

Subpart B – Current Good Manufacturing Practice

Subpart B addresses Good Manufacturing Practice (GMP), which has been used in all kinds of manufacturing for years. It encompasses the set of standards used to determine if a manufacturer is maintaining conditions and practices that are proven conducive to producing a safe food product. GMP helps to prevent contamination, mix-ups, deviations, failures, and errors, and is the foundation for any brewery's quality assurance program.[3] GMP is so important that chapter 3 is solely dedicated to it, where I look at each of the following sections in more detail:

§117.10	Personnel
§117.20	Plant and grounds
§117.35	Sanitary operations
§117.37	Sanitary facilities and controls
§117.40	Equipment and utensils
§117.80	Processes and controls
§117.93	Warehousing and distribution
§117.95	Holding and distribution of human food by-products for use as animal food
§117.110	Defect action levels

Subpart D – Modified Requirements

Subpart D states that the owner, operator, or agent in charge must register the brewery with the FDA. This can be done online and re-registration is every two years. This puts you on the FDA's radar, which is not necessarily a bad thing. For example, if there is a recall from another manufacturer in your area and the FDA believes you may use that material, they can reach out and let you know before it makes it into your production process.

Subpart E – Withdrawal of Qualified Facility Exemption

Subpart E outlines the processes the FDA can use to fine your facility, shut it down if dangerous or adulterated product makes it to the marketplace (this includes spent grains; *see* "Spent Grains FAQ" on p. 19), or ultimately withdraw part or all of your products from the marketplace. If you are doing your due diligence and complying with FSMA regulation, you should never have to review this section.

Subpart F – Requirements Applying to Records that Must Be Established and Maintained

Subpart F states that records must be kept and contain the actual values and observations obtained during monitoring and, as appropriate, during verification activities. Records should be as detailed as necessary and include:

- information adequate to identify the plant or facility (e.g., the name and, when necessary, the location of the plant or facility);
- the date and, when appropriate, the time of the activity documented;
- the signature or initials of the person performing the activity; and
- where appropriate, the identity of the product and the lot code, if any.

This sounds like a lot of information, but once you have standard operating procedures (SOPs) and training documentation in place you do not have to "write down everything." For example, if you have documented the SOP for cleaning/sanitizing a tank and you have documented that employee A has been trained on that SOP, then all that employee has to do each time is initial that they completed the task. We will cover SOPs in chapter 4.

---

3 "Good Manufacturing Practices for Craft Brewers," accessed September 15, 2018, https://www.brewersassociation.org/educational-publications/good-manufacturing-practices-for-craft-brewers/ [subscription required].

WHAT IF MY FACILITY IS A BREWPUB?

Brewpubs are themselves exempted from FSMA rules if they only sell directly to a customer, but they are required to comply with the FDA Food Code, which details food safety requirements within a food service environment. If a brewpub manufactures beer and sells the beer to other facilities, then it would need to comply with applicable FSMA rules. Depending on the size of operation, brewpubs may fall into the Qualified Facility status.[4]

HOW DO I COMPLY?

Do not let the idea of FSMA compliance overwhelm you. Let us break this down together. First, go online and make sure your facility is registered with the FDA (21 C.F.R. 1.225).

The overarching goal of the FSMA is to encourage manufacturers to take a proactive approach to producing safe food. So, your next step is to provide evidence that you are taking a proactive approach to food safety by writing and executing a GMP program and documenting its use. You must build and implement a GMP program and appoint a key person or small team at your brewery to oversee training. You can use your GMP program to outline this training.

Make sure to document all training. This will provide proof that a staff member is qualified to perform the task at hand. Once all staff members have received training appropriate for their assigned duties, document appropriate information when tasks are completed. For example, "John Smith cleaned, inspected, and sanitized Fermentor 1 on 10/12/2020." This documentation can help prove that you did the right thing.

In our brewery we like to say, "Write it down!" As our brewery grows, we document more and more digitally as the paperwork was becoming a hassle to keep up with and keep track of, but the documentation is still done regardless of the change from paper to digital. If training or a procedure is not documented, it did not happen in the eyes of the inspector. Protect your business. Although not explicitly mandated for breweries, it is recommended that as an extension of your GMP program you put a food safety plan in place.

GMP program + Food Safety Plan + Training + Documentation = FSMA Compliance

Of course, this is an oversimplification of things, but I want you to realize that FSMA compliance is not only required by law, it is also attainable at your brewery.

[4] "Food Safety Modernization Act FAQs for Brewers," Brewers Association [website], accessed September 15, 2018, https://www.brewersassociation.org/educational-publications/food-safety-modernization-act-faqs-for-brewers/ [subscription required].

2

GOOD MANUFACTURING PRACTICE

Quality Assurance and Quality Control (QA/QC) is not a stand-alone program, it is part of a larger quality system that includes Good Manufacturing Practice (GMP) and food safety programs. Luckily, there are many resources available to help you put together your quality system. The Quality Priority Pyramid (fig. 2.1) is a visual representation of six essential features and the general, prioritized order in which they should be put into practice to develop a solid quality program. Throughout this text I will use this pyramid as a general guideline to build our quality system.

While the implementation of the six tiers are prioritized, higher level tiers can also be addressed before a lower level tier is completely implemented. In practice, this is probably how it will be for you as you work toward creating a robust quality system.

Looking at the first step in the pyramid, you may be saying to yourself, "Manufacturing? What does this have to do with me? I am brewing, not manufacturing." Well, fundamentally, brewers are manufacturers. Brewers manipulate raw materials and through mechanical, chemical, and biological processes a final product is produced: beer. It is through this manufacturing lens that we will start to build a solid foundation for our QA/QC program.

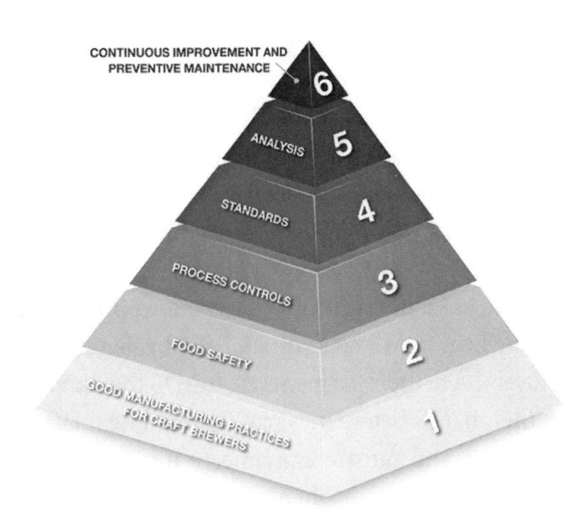

Figure 2.1. The Brewers Association's Quality Priority Pyramid, created by the BA Quality Subcommittee, is a visual guide of the six layers of quality that should be in place at every brewery. Image courtesy of Brewers Association.

WHAT IS GOOD MANUFACTURING PRACTICE?

Good Manufacturing Practice covers
Personnel
Building
Equipment
Processes

GMP is the set of standards used to determine if a manufacturer is maintaining practices set by federal, state, and county regulations.[1] At its core, GMP standards are intended to be the guidelines for safe handling, processing/manufacturing, packing, and holding of food products for human consumption. These GMP standards are required for every craft brewery and considered the minimum standard for what is necessary to make safe beer and food. You are encouraged to go beyond what the *Code of Federal Regulations* (C.F.R.) recommends for GMP.

At first, GMP was intimidating to me. After working with and developing a deeper understanding, I like to think of GMP as a set of rules for good housekeeping that ensure that your building (brewery and warehouse), equipment, and personnel are fit for beer production. Housekeeping is not intimidating, but it is something that must be done.

GMP sets the standards for the equipment used to make your beer. Essentially, the equipment must be cleanable. GMP addresses how waste and wastewater is removed from your facility. For example, do drains back up and contaminate your floors, your boots, then your hands that remove your boots and touch your beer? If this happens then your facility is not GMP compliant and the problem must be addressed, fixed, and documented. Notice I am not defining how it should be fixed. A construction crew may not be necessary if a critical control point in your procedure is identified and documented to stop the flooding.

Once your facility is up to code you need people to make the beer. GMP sets rules for the hygiene of your employees. Are their garments clean? Do they know to wash their hands before they handle anything related to production? You must communicate hygiene expectations and enforce them when necessary.

Now you have the facility and the people. What measures are you and your staff taking to ensure your cleaning processes are sufficient for producing safe beer? Are you checking that your ingredients are suitable for making safe beer? Putting standard operating procedures (SOP) in place and documenting the successful completion of tasks can address that.

Now that your beer is being made safely, your final GMP consideration is how you store and handle raw ingredients and finished product. Are you warehousing your raw materials and finished product safely? Addressing pest control issues, lot tracking, adhering to shelf life recommendations of ingredients, and using first in, first out (FIFO) policies will help with compliance.

It is important to note that the Food Safety Modernization Act (FSMA) specifically defines alcoholic beverages as food. Therefore, breweries are classified as food production facilities and are now under direct Food and Drug Administration (FDA) regulation (*see* chap. 1). Your brewery is legally required to write and implement a GMP program. If you are not in compliance with FDA regulations, your brewery can be subject to fines or worse.

The information that follows is intended to be a guideline to develop a small craft brewery's GMP program but does not guarantee that you will be in

GOOD MANUFACTURING PRACTICE VERSUS STANDARD OPERATING PROCEDURES

There is a difference between GMP and SOPs (also called sanitation standard operating procedures, SSOPs). As I mentioned before, GMP standards define the measures used to keep food sanitary and define general hygiene measures. They outline what you have to do, not necessarily how. The SOPs describe actual procedures that will be used to accomplish sanitary food handling and general hygiene practices. They will also document how monitoring and corrective actions will be accomplished and who is responsible. GMP standards and SOPs work together.

SOPs and their associated documentation (training documentation, checklists, and completion forms) are invaluable to your quality system. SOPs are the bridge that connects GMP to QA/QC. Without them we cannot get anything done. SOPs will standardize work that needs to be done, ensuring tasks are repeatedly completed successfully. Repeatability is the first step toward successful QA/QC. GMP consists of the fundamental tasks that need to be performed for your QA/QC plan, and the SOPs are the way you execute those tasks. SOPs will be discussed in detail in chapter 4.

[1] "Good Manufacturing Practices for Craft Brewers," Brewers Association (website), accessed September 15, 2018, https://www.brewersassociation.org/educational-publications/good-manufacturing-practices-for-craft-brewers/.

partial or full compliance with FDA, state, or local regulations. It is up to each brewery to ensure that they are in compliance. If you are still in the planning stages of your brewery, I highly recommend that you thoroughly research FDA, state, and local regulations and use your findings to inform equipment purchases and facility construction.

GMP REQUIREMENTS

GMP will serve as the foundation of your QA/QC program. We were all taught that cleaning is the first step in brewing. We even joke and call ourselves yeast janitors, which is not too far from the truth. GMP is the logical jumping-off point when building your quality system because GMP outlines the minimum sanitary standards that a food processing facility must meet. GMP includes personnel, plant and grounds, sanitary operations, sanitary facilities and controls, equipment and utensils, processes and controls, and warehousing and distribution. Without meeting these minimum standards set by GMP, you cannot guarantee a safe product.

To get your GMP manual started, familiarize yourself with C.F.R. title 21, chapter I, part 117, subpart B where the GMP requirements are listed.[2]

§117.10	Personnel
§117.20	Plant and grounds
§117.35	Sanitary operations
§117.37	Sanitary facilities and controls
§117.40	Equipment and utensils
§117.80	Processes and controls
§117.93	Warehousing and distribution
§117.95	Holding and distribution of human food by-products for use as animal food
§117.110	Defect action levels

These regulations are quite comprehensive. They are written to include all food manufacturing facilities so not everything will be relevant to your brewery. To help sort through this, I have highlighted some of the important points for breweries from each section. I encourage you to go through 21 C.F.R. 117.10–117.110 at some point as there may be things in there that are important to your specific facility that I have not included.

[2] 21 C.F.R. 117.10–117.110. You can access the current C.F.R. online at https://www.ecfr.gov/.

Personnel

If your craft brewery is anything like mine, there is very little automation present. This means there are a lot of points of contact between people and the beer, so there must be a minimum standard for cleanliness when handling any food product such as beer. The personnel portion of your GMP manual should address expectations relating to anything from your staffs' clothing and accessories, to hand washing, to general health concerns. Here are few areas your GMP manual should cover, with examples of what the brewery's policies are toward specific situations:

- Disease control
  - What to do if an employee is sick or injured, e.g., they are bleeding, have open sores
- Human cleanliness and safety
  - Clean clothes
  - Wearing of jewelry, e.g., unsecured stones must be removed because they are a choking hazard
  - When gloves should be worn, e.g., when in contact with food processing surfaces
  - Food, gum, and tobacco—it is prohibited to eat and/or chew when working in production areas.
  - Allergen cross-contact
- Hand washing
  - Clean hands are important to food safety—thoroughly wash your hands before entering production areas, after using the restroom, after eating or drinking, or anytime they become soiled.
  - Detailed hand-washing signage is required.

Plant and Grounds

The grounds about your brewery must be kept in a condition that will protect against the contamination of food. Adequate maintenance of grounds must include the following:

- Outside
  - Drainage, no standing water, and water drains away from building and production areas
  - Properly stored equipment; removal of litter and waste; and cutting of weeds or grass within the immediate vicinity of the plant that may constitute an attractant, breeding place, or harborage for pests.

    - If raw ingredients are stored outside, such as in a silo, ensuring that the silo is in working condition and will keep contaminants out.
- Inside
    - Facility constructed in a way that can be cleaned
    - Adequate lighting
    - Handwashing sinks must be independent from other sinks (e.g., bathroom or dishwashing sinks) and be near working areas.

Sanitary Operations

The brewery must be maintained in a clean and sanitary condition and must be kept in repair adequate to prevent food from becoming adulterated. *Adulterated* is a legal term used to refer to a food product that may have been produced, prepared, packed, or held under unsanitary conditions in which it may have become contaminated, or in which it may have been rendered diseased, unwholesome, or injurious to health.

- Cleaning and sanitizing of utensils and equipment must be conducted in a manner that protects against allergen cross-contact and against contamination of food, food contact surfaces, or food-packaging materials.
- Unnecessary toxic materials are not kept in the plant, only those that are used for operations such as cleaning and sanitizing are allowed.
- Be aware of non-food safe lubricants where there is the possibility of accidental inclusion.
- Pest control
    - Pests, large and small, can compromise the cleanliness of an otherwise sanitary facility. They transmit disease and track bacteria wherever they go. Work with a pest control service to plan how to keep pests out of your facility.
- Sanitation of both food contact and non-food contact surfaces

Sanitary Facilities and Controls

Sanitary facilities and controls outline conditions for process water in and wastewater out, as well as appropriately equipping personnel so they can carry out their duties in a sanitary manner.

- Water coming in must be adequate and safe for all processes, including cleaning and food contact and non-food contact surfaces.
- Wastewater disposal must be adequate to handle expected flow rates. There must be no drain back-ups that will contaminate food.
- Adequate hand-washing facilities
- Clean bathrooms
- Rubbish disposal

Equipment and Utensils

All equipment must be cleanable, inside and out.

- Equipment is to be installed to promote ease of cleaning.
- Equipment and utensils must be designed, constructed, and used appropriately to avoid the adulteration of food with lubricants, fuel, metal fragments, contaminated water, or any other contaminants.
- Instruments and controls used for measuring, regulating, or recording temperatures, pH, acidity, water activity, or other conditions that control or prevent the growth of undesirable microorganisms in food must be accurate and precise and adequately maintained, and adequate in number for their designated uses.
- Food contact surfaces must be in good repair and cleanable—no rust pitting or scratches.

Processes and Controls

Being a quality control person, Processes and Controls is my favorite part of GMP. It points out that brewers need to implement a quality system, pay attention to it, and maintain it. This is the heart of your quality control program.

- Appropriate quality control operations must be employed to ensure that food is suitable for human consumption and that food-packaging materials are safe and suitable.
- All operations in the manufacturing, processing, packing, and holding of food must be conducted in accordance with adequate sanitation principles.
- All contaminated or adulterated food will be rejected (do not try to dilute it out).
- Adequate precautions must be taken to ensure that production procedures do not contribute to allergen cross-contact and to contamination from any source (e.g., full cleaning of the brite tank after packaging peanut butter stout).

- Inspection of raw materials (I have found snail shells and bugs in grain before)
- Workflow must be able to be carried out in a sanitary manner.

SPENT GRAIN FAQs

Q: Can I donate/sell my spent grains to a local farmer?

A: Yes, brewers can send their spent grain out for animal feed if the following two conditions are met:

- You comply with good manufacturing practices.
- You are not further processing (e.g., drying) spent grain.

Q: Can I make bread, pizza dough, or another food product for human consumption with my spent grain?

A: That depends. If spent grain is processed further in any way and/or going into a human food product, the game changes. The spent grain then needs to be treated like food, which may subject your facility to parts of FSMA that a brewery would otherwise be exempt from (*see* fig. 1.1 on p. 11). These are inherently more stringent and will require extra time and resources to follow.

The FDA does have some exemptions for brewpubs and small businesses. Using spent grain in another food product can be done, but make sure you do your homework and do not inadvertently break an FSMA rule.

Warehousing and Distribution

Follow similar GMP principles when storing and transporting product and ingredients to and from the brewery.

- Storage and transportation of food must be under conditions that will protect against allergen cross-contact and against biological, chemical (including radiological), and physical contamination of food, as well as against deterioration of the food and the container.

Holding and Distribution of Human Food By-Products

Treat your spent grains with respect. These grains are going into our food system and in turn will make its way back on to yours or someone else's dinner plate.

Containers and equipment used to convey or hold human food by-products for use as animal food before distribution must be:

- designed, constructed of appropriate material, cleaned as necessary, and maintained to protect against the contamination of human food by-products for use as animal food;
- identified as such; and
- labeled so as not to be mixed up with trash and other potentially harmful substances.

Defect Action Limits

Defects are unavoidable but do everything in your power to prevent them.

- The manufacturer, processor, packer, and holder of food must at all times use quality control operations that reduce natural or unavoidable defects to the lowest level currently feasible.

Now, I know those nine sections are a lot to take in. But, when you think about it, they break down to the four areas I mentioned at the start:

Good Manufacturing Practice covers	
Personnel	Personnel are clean and healthy
Building	Building and grounds are clean
Equipment	Equipment is cleanable
Processes	Processes result in a safe product

Most likely your state-level department of agriculture inspection has already covered your process in this way. If you passed that inspection you are well on your way to having a successful GMP program. The next step is to put it in writing.

CULTURE OF FOOD SAFETY AND QUALITY FIRST

Before we even get started writing our GMP manual, let us get philosophical. Your company's culture is huge for your GMP program. For a program to take off, there needs to be buy-in from management. If there is push back, a reminder that a GMP program is required by law in order to be an FSMA-compliant business should be encouragement enough to get time and resources to start a program of your own.

The next step is to lead by example, setting in place a culture of food safety and quality first. Think about it, not only are you serving your beer to the public, but you are consuming it yourself and sharing it with friends and family. As a rule of thumb, if I come across any questionable food safety or quality issues, I think to myself, would I serve this to my family? If the answer is no, then the best decision is to take corrective action. This sentiment can be the seed that starts a culture of food safety and quality first.

WRITING A GMP MANUAL

We can philosophize about food safety culture all day, but we still have to put our GMP standards in writing for the program to work. I encourage you to use GMP as a personification of food safety culture. I mean, it is right there in the name "good." If you follow these practices, you are making good decisions. At first, you will want to keep your GMP manual simple, but you can and should update it as your process evolves and decisions get harder. Keep in mind that you are writing this as policy for your staff to follow, not just to meet the FDA requirements. Make it applicable to your facility, being specific as needed.

So, where to start? We just went through 21 C.F.R. 117 subpart B in detail, so I will start there. Let us use each section of subpart B as our table of contents:

Good Manufacturing Practice Brewery X Table of contents
Personnel Plant and grounds Sanitary operations Sanitary facilities and controls Equipment and utensils Processes and controls Warehousing and distribution Holding and distribution of human food by-products for use as animal food Defect action levels

Now for each heading write a paragraph that summarizes what you are paying attention to and, if applicable, what you are doing to control it. For example:

Plant and grounds
In general, the production plant and grounds are constructed and maintained in such a way as to prevent pests, mold or mildew, and contamination of ingredients in storage and beer in process. Brewery grounds are inspected daily and facilities, inside and out, are maintained according to Brewery X's Master Sanitation Schedule (MSS)

That wasn't so bad, was it? That paragraph clearly states your intention on complying with GMP and points to documentation (the Master Sanitation Schedule, MSS) on how you intend to keep up with it. If you do not have an MSS yet, do not worry—relax, have a craft brew—we will get you set up with one later in chapter x. For now, continue going down the headings and put together the guts of you GMP manual. Appendix A shows an example GMP manual outline.

Now that the hard part is over, we can start thinking about the supporting parts of our GMP manual. This is the easy part and should include:

- A title page including the author, version, date, and date effective
- The purpose and scope of the document
- Declaration of who oversees GMP
- What are you doing to meet these GMP standards?

<table>
<tr><td colspan="3">Brewery X
Good Manufacturing Practice</td></tr>
<tr><td>Author: Merritt Waldron</td><td rowspan="2">Approved by: Brewery X head brewer</td><td>Version 1.0</td></tr>
<tr><td>Date: 2/17/19</td><td>Effective: 3/1/19</td></tr>
</table>

Purpose

Brewery X is committed to producing beer of the highest quality and safety. To this end, Brewery X has developed, documented, and codified its manufacturing practices to ensure that all beer produced in our facility is wholesome, safe, and delicious.

Scope

This document outlines the framework for our GMP and Food Safety management programs. This document works in conjunction with our SOP/SSOPs, production logs, written policies, and training procedures to create a comprehensive food safety and QA/QC management system.

Food safety is the responsibility of each and every employee at Brewery X, and the management empowers every employee to halt production any time he or she sees a violation of our current GMP or notes a potential food safety concern not currently addressed in this plan.

The practices listed apply to all employees, visitors, and contractors. The rules listed are consistent with Good Manufacturing Practice regulations administered by the Food and Drug Administration (FDA).

GMP team (or representative)

The GMP/food safety team is the group with responsibility for the development, implementation, and management of the GMP program. The GMP team is comprised as follows:

- Director of Brewing Operations
- Head Brewer
- Director of Quality Control or assigned representative from the QA/QC team
- Plant Engineer or assigned representative from the Maintenance team

Ultimate responsibility of the GMP program lies with the Director of Brewing Operations.

GMP TEAM

In many small breweries the responsibilities of the GMP team may fall on the shoulders of one person. Do not let this hold you back from starting a GMP program. Once you have written a program, it becomes much easier to hand off tasks and change responsibilities as you grow and new hires fill the roles listed (and ones that are not listed).

It is helpful to have another set of eyes when developing your GMP program but this does not have to be anyone from the production team. Anybody, including a partner or tasting room/service staff, can help review GMP, share ideas, interpret policies, and point out missing items.

Putting these elements together, you will have a concise but effective GMP manual. If your brewery is ever visited by the FDA, or some other governing authority, the first thing they ask for could be this GMP document. Since you put in the work, documented your intentions, and you are pointing to the actions you are doing to comply, the inspector will be a happy camper. While the inspector may still find things that are out of compliance, they are likely to be lenient and advise you how to better your operation. Without any semblance of an organized GMP program, however, you will have no documentation or common ground to stand on with the inspector, who may issue a fine or worse—remember FSMA subpart E gives the FDA the right to initiate seizure of product or shut down your facility if they consider your beer adulterated.

SELF-AUDITING

A GMP manual is a living document, meaning it is constantly evolving as your processes evolve. As you and your team get more experienced at producing beer, you should improve your GMP manual as well. One of the best tactics to employ here is conducting a yearly FDA-style self-audit.

Head over to the Brewers Association website and download their GMP checklists.[3] These are comprehensive lists of yes or no questions relating to each GMP section. They even suggest some best practices to help you comply. The Master Brewers Association of the Americas (MBAA) also has many resources relating to GMP development and implementation. You can find them under "Good Brewing Practices" (GBPs) at https://www.mbaa.com.

I suggest printing these checklists out yearly, at least, and handing copies to each employee. Have the employees fill them out as objectively as possible and write down a qualifier when a field is not met. Once complete, compile the data into a document and come up with a list of improvements with levels of priority. Give yourself some realistic deadlines on completing projects brought up by the audit. Append this self-audit report to your GMP manual. This kind of documentation makes your GMP efforts traceable—*traceability* is key to GMP compliance. Here is an example of such a report:

3 https://www.brewersassociation.org/educational-publications/good-manufacturing-practices-for-craft-brewers/.

June 1, 2019	**Brewery X FDA self-audit report**	**Author:** Merritt Waldron

Introduction

Beginning April 20, 2019, Brewery X's GMP team initiated its first GMP self-audit based on form FDA 2966 in conjunction with the Brewer's Association GMP checklists based on that FDA form. The team asked all production employees to fill out an audit report, hoping to gather as much data from various perspectives as possible.

The GMP team collated the data, highlighting areas for improvement, and used this information to create an action plan for improving facilities, SOPs, and policies to better reflect Brewery X's GMP goals.

Results

Plant and grounds

- Some equipment is closer than 18" to walls.
- Some harborage for pests found. Pigeon roosts under eaves.
- T5 lighting fixtures on production floor do not have shields to protect against broken glass.

Equipment and utensils

- Grease seal on MLT rake motor leaks.

Sanitary facilities and controls

- Bathrooms are adequately cleaned but no schedule and log is in place.

Processes and controls

- Scale for can weights not checked for calibration.

June 1, 2019	Brewery X FDA self-audit report	Author: Merritt Waldron

Action Plan

Once the data was collated, the GMP team met and prioritized areas for improvement. Priority was ranked from 1–3 (or n/a) based on the team's assessment of the severity of the threat to food safety and the complexity of the solution. Additionally, the team outlined the identified solutions and assigned responsibility for oversight of each area of concern.

Priority 1

Constitutes (i) a substantial potential food safety hazard or (ii) is a lower hazard potential and is easy to implement. These items shall be dealt with ASAP with a target of 8/1/2019 to have all items addressed.

- Grease seal on MLT rake motor leaks.
  - Replace seal. John, Maintenance
- T5 lighting fixtures on production floor do not have shields to protect against broken glass.
  - Replace or fabricate shields for production floor lighting. John, Maintenance
- Bathrooms are adequately cleaned but no schedule and log is in place.
  - Put schedule and log in place. Janet, Head Brewer

Priority 2

Constitutes (i) a low-to-moderate potential hazard to food safety and is complicated to implement or (ii) a low potential hazard and is easy to implement. Target for all priority 2 issues to be addressed is 9/1/2019.

- Scale for can weights not checked for calibration.
  - Include can scale in instrument calibration plan. Merritt, Quality Control
- Some harborage for pests found. Pigeon roosts under eaves.
  - Install netting–contact landlord. John, Maintenance

Priority 3

Constitutes a minimal potential hazard to food safety and is extremely complicated or difficult to implement. Priority 3 issues will be assessed for potential resolution, and monitored for changes to SOPs, mechanicals, or facilities infrastructure. No target completion date.

- Some equipment is closer than 18" to walls.
  - Revisit placement upon brewery expansion. Janet, Head Brewer

Completion dates

Priority 1: 7/29/19
Priority 2: 9/24/19
Priority 3: not addressed

Only two chapters in and we are already through our first GMP audit. Not too bad at all. We are well on our way to building a solid foundation for our quality program.

As mentioned, it is inevitable that your processes evolve. When that happens, add things to your GMP manual that call out your improvements. If you write it down in the GMP you can hold yourself accountable for the new actions—better yet, you can empower someone else by giving them the responsibility. Be proud of your beer! Implementing GMP will instill a sense of pride and ownership in both the brewery's work and its standing in the community.

Example items to add to GMP manual
Training plan and documentation
SOP/SSOP documentation and record keeping
Traceability and record keeping
Evolving traceability and recall plans
Traceability audit records
Spoilage policies
Documented proof of vendors compliance with GMPs

3
FOOD SAFETY PROGRAM

Food safety is the second tier addressed in building our QA/QC program (*see* fig. 2.1, p. 15). In chapter 2, we looked at starting a GMP program for your brewery. GMP and food safety go hand in hand. The GMP program outlines the basic environmental and operating conditions of your brewery required to support your food safety program. The food safety program gets closer to the heart of production and looks at specific tasks.

A food safety program is a set of written documents based on well-known food safety principles that incorporates hazard analysis, critical control points (CCP), and a recall plan. It delineates the set of procedures to be followed for monitoring and implementing corrective actions in order to produce a safe and wholesome product.

Safety from what? Safety from the worst-case scenario: releasing an adulterated product that is hazardous to public health and safety. An event like this would trigger a recall or at least a voluntary market withdrawal. The priority of your food safety program is avoiding or minimizing potential for a recall. The principles embedded in your food safety program trickle down into the QC program, which also reduces the chances of a quality-based recall.

Recalls and withdrawals are expensive. Not only do you lose profits, but it also costs you money to pull back all the product and you lose the trust of your distribution partners, retailers, and consumers. Trust me, you do not want to wait for a recall or withdrawal event to start a QC system. I speak from experience—it was a voluntary market withdrawal that initiated the first version of a QC system during my time at Rising Tide Brewing Co.

SETTING UP A FOOD SAFETY PROGRAM

There is no one-size-fits-all food safety program. Your program will be specific to your facility and what you are making. Here we will discuss methods used to develop a typical food safety program. Many of these methods are directly transferable to developing a QA/QC program. The method we will use consists of four steps.

1. Identify potential hazards. A hazard is any biological, chemical, or physical agent that has the potential to cause illness or injury.
2. Identify critical control points. A critical control point (CCP) is a process step or procedure in the brewing process at which control can be applied and is essential to prevent or eliminate a food safety hazard or reduce a hazard to an acceptable level.

3. Define conditions to manage the hazard. These will inform your control methods in which you measure and document those conditions.
4. Define critical control limits. A critical control limit is the maximum or minimum value, or combination of values, to which any biological, chemical, or physical agent must be controlled to significantly minimize or prevent a hazard. This limit will trigger a corrective action that is predefined in order to swiftly prevent the food safety issue from getting worse.

Let us go through each of the steps in closer detail to get a better understanding of what to look for.

Identifying Potential Food Safety Hazards

There are three main sources of hazards found around the brewery, which are physical hazards, chemical hazards, and biological hazards. The following is a partial list of each hazard commonly found in the brewery. It is not meant to be a complete list but examples of things to look out for when developing your food safety program.

Physical Hazards

- Foreign object inclusion: glass, plastic, metal, or wood pieces of any size found in finished, sealed product.
- Package explosion: bottle, can, or keg overpressurization and failure. This is an important issue in the craft brewing industry right now—see the sidebar "Package Refermentation" on pages 27–28.

Chemical Hazards

- Chemicals: can become part of food without being intentionally added. For example, residual chemical can be left in lines or on tank walls due to improper rinsing; or, if chemical dosing is not carefully measured, a "no-rinse" sanitizer can be pushed out of the range for its no-rinse application, thus requiring a rinse.
- Toxicants: toxins produced by living organisms, microbes, plants, or animals.
- Allergens: substances that cause abnormal reactions in the immune system. Allergic reactions can range from a slight rash to severe injury and death.
- For other food products, the FDA requires that products including the following ingredients be labeled: milk, eggs, fish, crustacean shellfish, tree nuts, peanuts, wheat, and soybeans. At the time of writing, it is not required for craft brewers to include this information on a label. However, it is a good idea to get in the habit of including this information somewhere on the label, especially if it is not disclosed in the style description, such as wheat beer, or milk or oyster stout.

Biological Hazards

- Pathogenic bacteria, viruses, and parasites: can be introduced into food product via ingredients and raw materials, from food processing equipment and environments used to make the final product, and from people handling, harvesting, or processing.

CAN PATHOGENS LIVE IN BEER?

While the inherent characteristics of beer (pH <4.6, alcohol, antimicrobial activity of hops) help protect against pathogenic bacteria, this does not guarantee that pathogenic bacteria cannot exist, survive, or thrive in beer. As brewers continue to innovate with higher pH, lower ABV, non-traditional ingredients, additions post-boil, and copious amounts of fruit, the protections we take for granted can be weakened or lost. Do not be the first brewer to push the boundary too far and create an environment for pathogenic bacteria. Keep in mind these bacteria can thrive on packaging material and raw ingredients. You can use the same strategies that mitigate contamination by wild yeast and bacteria to also reduce the risk of pathogenic bacteria contamination.

PACKAGE REFERMENTATION

As I write this, one issue currently plaguing the craft brewing industry is an abundance of market withdrawals that stem from package failure. Homebrewers would call these "bottle bombs" and, in many cases, it is due to overpressurization of the vessel leading to bottles and cans on the shelf exploding. This is unacceptable for a professional brewer and is reasonably easy to avoid. Let us take a look at this issue in closer detail.

WHY PACKAGE REFERMENTATION IS A PROBLEM

First, overpressurized vessels are a serious safety hazard to your consumers, retail partners, distributors, and staff. Exploding bottles and cans can project shrapnel in every direction. I have seen pull tabs open upward with the slightest touch, beer burst out the sidewall of cans, and glass bottles pop and crumble. I have even heard a tale of a burst bottle in a retail store that pierced the chip bags in a display across the aisle. This is very serious and scary stuff. The first step in preventing this from happening is understanding the relationship between CO_2 and fermentable extract.

As yeast metabolizes fermentable extract two main products are produced, ethanol and CO_2. In a closed vessel, such as a bottle, keg, or can, the CO_2 has nowhere to go, thus, the internal pressure of the package increases. The amount of CO_2 produced is dependent on the amount of fermentable extract left in the vessel. For example, if you package a beer with yeast and the beer is 1°P (1.004 SG) of fermentable extract above true terminal gravity, over time about 2.50 volumes of extra CO_2 will be produced. The CO_2 produced in this case will be in addition to the CO_2 already in package, so if you packaged at 2.55 volumes this brings you up to approximately 5.05 volumes in package. Depending on the pressure limitations of your package that level of pressurization could put you and others in danger.

CO_2 volumes as a function of residual extract
1°P (1.004 SG) ≈ 2.5 vol CO_2
0.1°P (1.0005 SG) ≈ 0.25 vol CO_2

At this point it is important to state storing beer cold is not an acceptable quality measure. It is absolutely the responsibility of the brewer to make sure their product is shelf stable.

CAUSES OF PACKAGE REFERMENTATION

Let us look at some possible causes of package refermentation a little closer in order to understand how to prevent it from happening.

Unfinished beer in package. The root cause of beer that continues to ferment in package is simple: for one reason or another the true terminal gravity has not been reached. It can be caused if the brewer fails to account for or spot "lazy attenuation" on the part of the yeast. Essentially, you are unintentionally bottle conditioning your beer. The yeast that remains in package continues to ferment, producing CO_2. Because CO_2 has already been introduced at packaging, any additional CO_2 can easily push the package beyond its pressure limits.

Hop "creep." Hops have been shown to contain small quantities of the enzymes amyloglucosidase, α-amylase, and β-amylase.* These are the same enzymes that break down complex starches into fermentable sugars in the mash and they will have the same effect on any remaining starches in beer after initial fermentation is finished. The increase in fermentables that causes refermentation is commonly termed hop creep. It usually occurs due to high dry-hop loads and insufficient refermentation time. Higher levels of dry hops contribute more enzymes, which leads to higher levels of fermentable sugars being introduced into the finished beer.

* Perkins, J. Shellhammer, T. Bodah, Z. 2017. "Unintended Over-Attenuation from Dry Hopping Beers" Presentation at Craft Brewers Conference. Washington, D.C.

Late sugar additions. Any ingredient that contains fermentable sugar, such as fruit, honey, molasses, or syrup must be allowed sufficient contact time with the yeast to allow it to completely ferment out.

Contamination with wild or domestic yeast. In-package fermentation can happen if a domestic yeast strain with higher attenuative properties or a wild yeast strain with unknown attenuative properties is included in package at some point in your process. This may look like lazy attenuation, except the source is not the yeast strain you think it is.

POSSIBLE ROOT CAUSES OF UNINTENDED FERMENTATION IN PACKAGE

Cause	Mitigation strategies
Lazy attenuation of yeast	Let the yeast finish. If it is taking longer than usual for your strain, look into yeast health (viability, vitality) and pitching practices (temp, O_2, cell load) Some strains have unique quirks. For example, certain *Brettanomyces* and saison strains are known to be lazy attenuators, where they appear to be done but have just slowed down and can take several more weeks to truly finish. Get to know each new yeast strain you bring in by piloting and mimicking production conditions as best as you can
Hop creep	Monitor refermentation and allow extra time after dry hopping Pilot new hop recipes and techniques mimicking production conditions. There are many factors to consider: Hop variety and crop year Hop contact time Residual extract Temperature and residual live yeast
Late sugar additions	Contact time with yeast at appropriate fermentation temperatures
Contamination with wild or domestic yeast	Rely on (or start up) a consistent microbiology program to detect contamination (see chapter 14) Use rigorous and validated sanitation procedures

VESSEL LIMITATIONS: TEMPERATURE AND PRESSURE

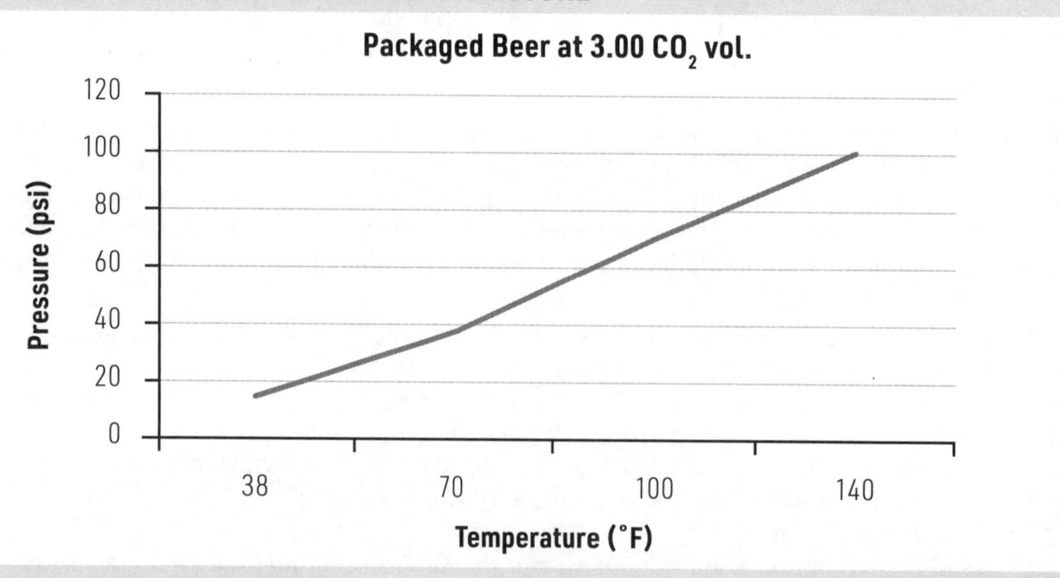

Reproduced by permission from J. Perkins and Z. Bodah, "101 Ways to Blow Up a Bottle/Can, and How to Avoid It," seminar presented at Craft Brewers Conference, Washington, D.C., 2017.

Critical Control Points and General Control Methods

Critical control points are a process step or procedure in the brewing process at which control can be applied. For our food safety plan, we will want to focus on the points in which product is exposed to a hazard or itself can be a hazard. Be as complete as you can when identifying critical control points. Look at each process step with regards to physical, chemical, and biological hazards. Think about the worst-case scenario. What if something fell in? Can bacteria grow here? You can also look to other recalls to inform your control points. What was the recall issue? What could the manufacturer have done to prevent that? Can I measure and manage that same control point?

After identifying critical control points and flagging a potential hazard, it is time to define and implement a control method. For example, a choking hazard may have found its way into a can (control point: hazard in empty can), so you will visually inspect every can (control method: visual inspection). Next you should ask, does this method control the hazard? Yes, if the operator is paying attention. Visual inspection of every can may be a fine control method if canning by hand, but you must ensure staff are trained on this and it is added to the SOP. Some other examples are shown in table 3.1.

Now we must examine the feasibility of monitoring the control. As we start to can faster or automate the canning system, checking every can will not be feasible, so you should adjust the control method. Your control point (hazard in empty container) stays the same, but now we are going to invert all cans through a twist rinse (control method: invert and rinse) to remove the hazard. You should test this control method and document the results. To test the twist rinse, you can place common choking hazards (e.g., plastic labels, pallet wood pieces, even bugs) into the cans beforehand to test the efficacy of the can rinser and document the results.

Finally, there is such a thing as too many control points. If a hazard is not reasonably likely to pose illness or injury, it likely does not have to be monitored. Too many extraneous controls can result in a system that cannot be effectively managed with available resources.

TABLE 3.1 COMMON CONTROL METHODS FOR BIOLOGICAL, CHEMICAL, AND PHYSICAL HAZARDS

Physical hazard	Broken glass policy based on documented hazard analysis Filtering or separation Metal detection or X-ray devices
Chemical contamination	Validated sanitation controls Process controls relating to chemicals Allergen labeling Due diligence ingredient sourcing Cross contamination managed through GMP
Biological contamination	Good personnel hygiene Good sanitation Cross contamination managed through GMP Kill step Heat Acid Antimicrobial chemicals Irradiation Pressure

Critical Control Limits

Critical control limits are an important part of your food safety program. The critical control limit is the maximum or minimum value, or combination of values, within which any biological, chemical, or physical agent must be controlled to significantly minimize or prevent a hazard. Control limits should meet three requirements: they must be achievable, effective as an indicator of the CCP, and monitored at regular intervals. Some examples of control limits are time, temperature, pH, flow rate, and belt speed through pasteurizer. These control limits are typically process dependent. The selection of the best critical control limit for a CCP can be informed by scientific research, FDA guidelines, or by practicality and experience.

A good example of a critical control limit that should be checked daily is the pH of your rinse water after clean-in-place (CIP) of a tank that has been cleaned with caustic. Using pH as the indicator, you can show that your high-alkaline cleaner (caustic) has been removed from the walls of your vessel. If you know the pH of your rinse water source, you can easily identify a reasonable limit or target for your final rinse. Let us walk through a step-by-step example.

Step 1. Identify potential hazard	Chemical inclusion in product
Step 2. Identify critical control point(s)	Rinse post-CIP cycle
Step 3. Define conditions to manage the hazard and implement a control method	Ensure pH on rinse water samples fall between critical values. Inspect tanks and fill out CIP sanitation log
Step 4. Define critical limits	pH is to fall between 6.0 and 7.5 at final rinse

ASSEMBLING A GMP/FOOD SAFETY TEAM

Armed with a general idea of what hazards you are looking for and ready to identify control points, you should put together your GMP/food safety team. This team will be responsible for developing documents and policy surrounding GMP and food safety programs. The team will oversee implementation, ensure training on policy, and will maintain the programs with regular audits.

These responsibilities could fall to one staff member. If you are lucky enough to have a few staff members, assemble a team of people with different specialties and experiences. Again, management buy-in is key, but it could be useful to have folks from each department: QA/QC, brewing, cellar, packaging, and maybe even tasting room (or front of house). This spreads the workload and empowers the folks who are getting this import work done. Just be sure that responsibilities are clearly laid out in writing (see below). As always, time is limited in a brewery setting, so start with the CCPs that mitigate the biggest risks first, such as accidental chemical inclusions, glass inclusion or similar physical hazards, and overpressurization of vessels due to continued fermentation. After those are complete, chip away at the less urgent tasks.

WHAT IS HACCP?

HACCP stands for Hazard Analysis and Critical Control Point (HACCP). It is a scientific, systematic, and preventive approach to identifying, analyzing, and controlling significant food safety hazards within a food production system from raw material acquisition through to consumption of a finished product.

The HACCP system has two major components. The "HA" in HACCP represents the logic of the hazard analysis, identifying the where and how of hazards. The "CCP" represents the critical control points that provide the control of the process and the proof of the control.*

HACCP comprises seven principles,† which can be applied to a food safety program.

Principle 1: Conduct a hazard analysis

Principle 2: Determine critical control points (CCPs)

Principle 3: Establish critical limits

Principle 4: Establish monitoring procedures

Principle 5: Establish corrective actions

Principle 6: Establish verification procedures

Principle 7: Establish record-keeping and documentation procedures

* "Introduction to HACCP Training," University of Nebraska-Lincoln (website), accessed September 14, 2019, https://food.unl.edu/introduction-haccp-training.

† "HACCP Principles and Application Guidelines," U.S. Food and Drug Administration (website), accessed September 14, 2019, https://www.fda.gov/food/hazard-analysis-critical-control-point-haccp/haccp-principles-application-guidelines.

While an HACCP (see sidebar) program is not mandated at time of writing, I am applying HACCP principles here to develop your food safety program. When you are ready to take it to the next step and marry your food safety program with an HACCP program, there are many HACCP resources provided by trade organizations, such as the Brewers Association (BA) and the Master Brewers Association of America (MBAA). These will help you further your understanding. There are also HACCP and food safety resources offered by local universities, such as cooperative extension programs that provide free education services to local communities. HACCP certification can be a great continuing education opportunity for a valued staff member.

Put It in Writing

Now that the gang's all here, it is time to start putting your plan in writing. It is helpful to come up with a flowchart of your entire process. This will help keep the team organized and it is also helpful to show an inspector when they show up. Once the flowchart is done, you should devise a template for your documentation. As with the GMP manual, the BA has put out templates for food safety that are intentionally customizable. Each template focuses on a particular type of hazard and includes some control methods that can be implemented. With your templates in hand, try to identify if there is a hazard present for each step of the process. When you come across one, fill out the field in the template that answers the four questions from our method above:

1. Identify potential hazards (biological, chemical, or physical)
2. Identify CCPs
3. Define conditions to manage the hazard and implement a control method
4. Define critical limits and corrective actions

This can be a lot to take on, so it is helpful to split it up among the GMP/food safety team. Start simple and get more complex as you grow. The food safety program is another living document and will get better as you go on. You may also come across things you are not sure of. Do not worry, there are plenty of resources out there for you to consider, fo example:

- Employees: they know the process and will implement the plan.
- Suppliers and buyers: they know the raw materials and should be able to help identify any hazards associated with it.
- University specialists: many universities have a public safety or food safety program as part of their cooperative extension programs.
- Other breweries: consider especially breweries that are large and successful. See what they are doing—chances are they are doing something right.
- Trade associations: like the BA, MBAA, and American Society of Brewing Chemists (ASBC).
- Consultants and auditors: use these resources if there is something beyond your expertise. It is OK to not know things—do not be afraid to ask for help!

MONITORING PRINCIPLES

With your food safety program written, it is time to put it to work. Monitoring is where the work in maintaining food safety gets done. For the monitoring systems to be successful, you need to look at four main elements:

- What measurements or observations will be used to monitor?
- How will the monitoring be conducted?
- How frequently will monitoring be carried out?
- Who will conduct the monitoring?

Each operator should be able to take critical measurements and determine if they are within critical limits. This is where training and SOPs are going to be crucial. Each CCP should be mentioned in the SOP and training should be built around it. This helps ensure that the operator understands the process controls and their importance in keeping your beer safe, wholesome, and delicious.

RECALL PLANNING

In the event that adulterated food makes it to market or product out of specification leaves the plant, you should have a written recall plan in place. A written recall plan is an essential piece of your food safety program. If your CCPs are properly identified and controlled, you should never have to use the recall plan. If something does happen, you will be glad you have it. The recall plan will be your road map to get through a difficult time.

Types of Recall

Typically, a recall or withdrawal will fall under one of three broad categories. Voluntary market withdrawal is by far the most common kind of recall seen in the brewing industry. The next categories are rare in the brewing industry, but they are mentioned here for sake of completeness.

- Voluntary market withdrawal, for reasons such as out-of-specification beer, does not require regulatory notification except in a case where the recall is expected to be large enough to warrant requesting excise tax relief. In such a case, TTB must be notified.
- Recall initiated based on regulatory action by the FDA or TTB. In such a case, FDA and TTB will both be involved from the very beginning as initiators of the recall event.
- Recall for adulterated or mislabeled product, or other food-safety reason, as identified by the brewer. Such a recall should trigger the brewery to contact the TTB and FDA as soon as it is determined that a food-safety hazard exists.

The TTB offers guidance to alcoholic beverage producers in its Industry Circular 2017-4, "Voluntary Alcohol Beverage Recalls."

When developing your recall strategy, you should consider several factors. These factors may include the results of an FDA health hazard evaluation, the type of product, the usage patterns of the product, the ease in identifying the product, the degree to which the product's non-compliance with the law is obvious to the consumer, and the degree to which the product remains in the marketplace. Further, your recall strategy should address the following elements regarding the conduct of the recall: the need for publicity, the scope of the recall, and a measurement of effectiveness. Depending on the circumstances, recalls may involve wholesalers, importers, distributors, retailers, and consumers. The scope of the recall is determined by the severity of the problem or the circumstance that requires a removal from commerce. A recall decision does not depend solely on the health risk of the product. Adulterated products and mislabeled products where no health hazard exists are still in violation of the law, and voluntary recall may still be appropriate. (Industry Circular 2017-4 [7], last updated February 9, 2018, https://www.ttb.gov/industry-circulars/ttb-industry-circulars-17-4)

DATE LOT CODING

When packaging beer, it is in the best interest of brewers, distributors, importers, and consumers that all beer from small and independent craft brewers be identifiable by some form of a date or lot code.* It is the responsibility of all brewers to ensure that their products are clearly coded for traceability. If there is a reason to initiate a withdrawal or recall, a date code system is the most effective mechanism to get that beer off the shelves.

* "Date Lot Coding," Industry Updates, Brewers Association (website), February 1, 2019, https://www.brewersassociation.org/brewing-industry-updates/date-lot-coding/.

Initiating a Recall

The following points highlight the general approach you should take in the event that your brewery initiates a recall. Refer to appendix B for a more detailed checklist of what to include in your recall planning and procedures.

- Be the first to get the story out—that way you control the narrative.
- Stick to the script!
- A recall is an opportunity to prove that you are trustworthy and responsive—take advantage of that opportunity.
- Acknowledge the issue and take responsibility for it! Do not hide it or try to place blame elsewhere.
- Give all the facts, do not try to hide anything!
- Be open and honest about potential hazards/health risks.
- Communicate a commitment to ensuring public safety.

Mock Recalls

With your recall plan in place, it is a good idea to check and see if it works. If possible, this should be done before you actually have to use your recall plan for real. A mock recall is an effective exercise that will prepare your brewery for a recall and it should be conducted at least annually. The goal of a mock recall is to check if your brewery can effectively trace and retrieve its product from the market.

A mock recall will help test traceability and record-keeping practices and discover if any part of the recall plan is weak, problematic, or just plain broken. To get started, choose a product to recall. Start with a safety CCP you identified in your food safety program that is likely to trigger a recall if it were to fall undetected outside of the critical control limits. Once the product is chosen, assemble your recall team; this may be a cross departmental team depending on your brewery. The recall team should include the persons in charge of making the decision to recall (e.g., the director of brewing or quality), the sales or warehouse team who are in charge of getting the beer back, the CFO or accounting team members who will monitor the cost of refunds, and someone in charge of communicating to the public and your distributors. Finally, execute the mock recall by following your recall plan step by step. Document all actions conducted, deviations from the plan, how long it took you, how much product was "recovered," and if there are any problems following the plan as written.

Upon completion, make the time to reflect on how things went. Be honest and answer the following questions:

- Was the recall team able to quickly reach a decision regarding the recall?
- How difficult was it to:
    - trace the recalled product,
    - gather the information necessary to activate the recall,
    - contact regulatory agencies,
    - prepare documents for media?
    - Maintain a log of activities?

Using these insights, the recall plan should be modified to better reflect the actual steps employed and include details that were found to have been overlooked during the mock exercise.

Food safety is everyone's responsibility. Writing down and implementing a food safety program shows your staff and potential inspectors that you take it seriously. It introduces thoughtful use of control points and observations, a concept that will come up repeatedly while planning for quality control. A food safety program helps avoid indecision over how to handle a product or process whose controls are over the set critical limits. A food safety program is a critical step in building a robust and effective quality system in your brewery.

Things to Add to Your Food Safety Program
Conduct a mock recall to audit your program on a regular schedule
Audit anytime you add a new step or process
Create a flowchart of the production process (from ingredients to packaging and beyond)
Establish consumer and internal product complaint records once food safety program is established

RESOURCES

For more information, these resources can be helpful:

Brewers Association
https://www.brewersassociation.org/

Food Safety Preventative Controls Alliance (FSPCA)
https://www.ifsh.iit.edu/fspca

Food and Drug Administration (FDA), *Bad Bug Book: Foodborne Pathogenic Microorganisms and Natural Toxins,* 2nd edition
https://www.fda.gov/food/foodborne-pathogens/bad-bug-book-second-edition

4

STANDARD OPERATING PROCEDURES

If you have been around or working in the brewing industry for any amount of time, you have probably heard the phrase "S-O-P" numerous times. This is short for standard operating procedure. On the surface, an SOP is a simple set of instructions for a specific task. But when you dig deeper, SOPs are much more than that. They are a powerful tool that allows you to accomplish great things around the brewery in regard to your quality system. If you are following along with the Quality Priority Pyramid from chapter 2 (p. 15), you may have noticed we bypassed the third tier and jumped straight to the standards tier. SOPs are the perfect example of how you can work at each level of the pyramid simultaneously because each level supports the others. By introducing SOPs here, you can use them to put your GMP and food safety programs into action (tiers 1 and 2), as well as prepare yourself for further process controls (tier 3) and the execution of CCPs and QCPs, which we will discuss later.

WRITING A STANDARD OPERATING PROCEDURE

While SOPs are written sets of instructions for specific tasks, how they are written is up to you. The non-negotiable part of that last sentence is that SOPs must be written down. When they are not written down, the accomplishment of tasks in your brewery turns into a game of telephone. While the first person you show may accomplish the task the way you intended, if training continues to develop through word-of-mouth you can end up with the last person accomplishing the task with a completely different outcome. This is exactly what we are trying to avoid by putting in place our quality system. A written SOP will mitigate this problem by creating a fixed reference point for the work to be done. Deviations from your target can be analyzed more easily when the work is done using a consistent and agreed-upon process described in the SOP.

Level of Detail

When it comes to writing an SOP, there are two common schools of thought. The first is to produce a basic outline of the critical steps and control points. Let us look at an example of an SOP for an equilibrated dissolved oxygen (DO) reading taken with a DO meter that we would use to later calculate total packaged oxygen (TPO) in a can.[1]

---

[1] Jamie Floyd, Campbell Morrissy, Ben Parsons, and Merritt Waldron, "Writing it Down! SOPs The Foundation to Any Quality Program" (presentation, Craft Brewers Conference, Denver, CO, April 10, 2019).

Canning Run Equilibrated DO Samples
1. Pull can from respective fill head. 2. Weigh can (target 494 g ±7 g) and record weight. 3. Shake can for 1 minute. 4. Place can in piercer and engage. 5. Rinse until DO stabilizes and press "play" to record. 6. Record DO (target <75 ppb), CO_2 (target <0.10 vol drop from tank), and temperature

This is straight and to the point. Steps are presented in a logical order. Control points are indicated by parameter targets and tolerances. And the task can be accomplished reliably.

The second school of thought is to write an incredibly detailed how-to instruction guide. Using the same process as an example, let us take a look.[2]

Canning Run Equilibrated DO Samples
1. Pull can from respective fill head. 2. Weigh can (target 494 g ±7 g) on Ohaus CS200 (for calibration steps see SOP 04B-001 Scale Calibration). 3. Shake can for 1 minute, use timer on canning desk to track time. 4. Place can upside down in piercer, ensuring lip is firmly in recession. 5. Lock can in place with large thumb screw. Ensure height gauge is fully covered. 6. Firmly push cover down, locking into place. Occasionally, you will need to use your left hand to put pressure on cover to ensure it is locked. 7. Pierce can by pulling black lever toward you, an audible hiss will be heard. 8. Slowly drop stem into can until it touches the bottom of the can. Careful, too forceful of a drop can pierce through the top of the can, causing a mess. 9. Back stem out ½" and lock into place by twisting thumb screw. 10. Press rinse button, ensuring drain hose is clipped into a waste bucket. 11. Wait until DO stabilizes (±10 ppb) and press play to record. 12. Log DO (target <75 ppb), CO_2 (target <0.10 drop from tank), and temperature on canning tracking sheet.

Another perfectly good SOP. Again, steps are presented in a logical order. Control points are indicated by parameter targets and tolerances.

In practice, you will likely elect to use both schools of thought when writing your SOPs. For most tasks, I tend to use the basic outline with critical steps and control points. So, for the example task above, I would elect to put the basic version in my SOP binder. There are two reasons why I choose this approach: first, during training I would include instruction on the ancillary tasks and anecdotal information, such as where the timer lives, tighten the thumb screw, and listen for the hiss; second, it is much easier to read and reference the document for control points after the initial training. When it comes to potentially dangerous or super critical tasks, however, the very detailed how-to instruction guide is the only way to go.

So far, I have only glossed over training on the SOP. For basic or detailed SOPs to work as you intend, some serious time should be spent on training. I firmly believe you must do hands-on training for a task at least five times before you understand it and can work from the SOP. For tasks that may be dangerous make sure that trainees have a complete understanding of any associated risks and how they can be avoided. Employee safety should be at the front of your mind whenever writing or training on SOPs.

Supporting Components

While the list of critical steps and control points are the heart of the SOP, there are several supporting components that will unlock the full potential of this tool for your quality system. As I just mentioned, employee safety should be at the front of your mind whenever writing SOPs—your first supporting component is a hazard assessment of the SOP.

Hazard assessment. With your new SOP in hand, go through each step and identify any potential hazards that might exist. This includes chemicals used, where to find the safety data sheet (SDS), and any lockout/tagout (LOTO) procedures. If a potential hazard exists, determine its likelihood and severity. Put in place a means to eliminate or restrict the hazard. If necessary, rewrite that step in a fashion that allows staff to complete it in the safest way possible. Finally, at the beginning of the SOP, call out that safety concern and equipment necessary, such as personal protective equipment (PPE), to mitigate the risk. For a comprehensive look at hazard

[2] Floyd et al., "Writing it Down!"

assessment, look up the Brewers Association's *Hazard Assessment Principles*.[3]

Personal protective equipment (PPE). Include a list of PPE needed at the very beginning of the document. Operators should be prepared in advance for any potential hazards by referring to the SOP's hazard assessment (see previous paragraph). Keep in mind PPE is not a solution to correcting a hazard, but it is the last line of defense. Reinforce the importance of PPE during initial training and call out folks for not using PPE when it is listed.

List of equipment needed. The list of equipment needed should be reserved for items unique to the task, such as the appropriately sized pump, a certain length of hose, or the appropriate measurement instrument. You can be super detailed here and include things like the appropriate number of butterfly valve and tri clamps, but I find this to be a bit clumsy and hurts the readability of the SOP.

Order of operations. This was covered in the previous section (p. 36). Think carefully about the level of detail necessary so that staff can perform the task safely when following the SOP.

Control points. Call out your CCPs (see chap. 3, p. 30) or your quality control points (QCPs; addressed in chap. 6, p. 53). List the target value and tolerance developed when you set up your standards.

Tracking and traceability. If it is not written down, it did not get done. Since you have CCPs and QCPs in place, you will want to document and analyze those control points to verify the process is within the control parameters and working as expected. In addition to calling out your control points, make it clear where they are to be recorded. This creates a paper trail and documents that the SOP is being used. This can come in handy when troubleshooting a process or when an auditor comes in for an inspection. You can point out with confidence the task was accomplished, using this SOP, resulting in this outcome.

Pictorial aids. A picture is worth a thousand words. In the age of smart phones, it is easy to snap a photo and drop it into your document. If you are having trouble describing a set-up or procedure for a task, consider adding a photo or diagram of that step. Be wary of overcomplication: if you have to rely on a lot of photos, you may have an overly complicated process or set-up. Consider streamlining for efficiency's sake. Another great visual aid to consider is a flowchart that demonstrates the steps in the appropriate order to complete the task at hand. This is a terrific way to show an overview of the entire SOP, especially if it is a longer task.

Organizational details or updates. Include the purpose and scope of the SOP and where it is applicable. Apply a unique SOP number or code and a version number; also include a quick synopsis of changes from the prior version. These tracking elements come in handy as your SOP binder gets larger and more complicated. Keeping SOPs organized is imperative and doing so helps ensure everyone is working from the same, up-to-date SOP.

Table of contents. Use a table of contents when necessary if a document consists of multiple tasks to be completed in order.

Putting all these supporting components together turns a simple list of tasks into the ultimate QA/QC tool. They allow you to get jobs done safely and without confusion or argument. Gathering data becomes uniform and standardized. This aids in troubleshooting and analysis when the time comes.

Using a Template

So that was a lot to remember. How will you make sure you and your team are writing SOPs that contain all this information in the right order? This is where a template comes in handy, especially when there is more than one author. A template will force authors to include all the supporting components mentioned above. It facilitates interpretation and creates a company-wide standard for training and cross-training. The template you choose is completely up to you, but if you want help from staff with writing and updating SOPs, make your template in a software program that is easy to use. An example of a template layout is giving in Appendix C.

[3] Brewers Association Safety Subcommittee. *Best Management Practice (BMP) for the Development of Safety Programs in Breweries Volume 1: Hazard Assessment Principles*, Brewers Association (website), accessed August 3, 2018, https://www.brewersassociation.org/educational-publications/hazard-assessment-principles/.

IMPLEMENTATION, TRAINING, AND UPKEEP

When looking for a place to start writing and implementing SOPs, you should consider prioritizing safety, prerequisite programs, and then quality. I know it seems weird to hear that from the person writing a quality book, but I genuinely believe that quality beer cannot exist without these prerequisites.

Make sure SOPs are easily accessible. This could be in a binder or a shared file. If SOPs are digital, make sure only qualified personnel are able to make changes. Start rolling out new SOPs slowly so the staff is not overwhelmed. After all, quality beer is not a sprint but a marathon.

Before training, walk through the SOP step-by-step to ensure the process order makes sense and that nothing was missed in the hazard assessment. It will help to do this with a trusted operator or manager, as they may point out something that was missed. Keep a look out for steps that could negatively affect quality in any way. If the SOP is deemed ready, start the training process.

Training is serious business—all operators should understand the fundamental principles behind the SOP before being authorized to perform the task. Keeping a log of training sessions associated with each SOP is recommended. How you handle this is up to you. I have included a training log sheet in the example template in appendix C. Train with the SOP in hand and go step-by-step. Emphasize the importance of the SOP to the trainee. Once you are confident they can accomplish the task according to the SOP, have the trainee train the trainer. Ask the trainee questions along the way to encourage understanding of the task at hand. Once you are satisfied with their understanding and ability, have the trainee sign the training sheet acknowledging that they are able to perform the task without supervision. Not only does this emphasize the importance of the SOP, it also provides accountability. While this may seem very formal, it is essential to achieve consistency in your quality program. Remember the game of telephone.

To make this formal process work, you should keep an open mind to operators' needs as well. SOPs are living documents. As your process evolves, so will your SOPs. Be flexible, and update the SOP on the fly if needed. The "because I said so" approach does not really work in the adult world. Encourage and empower operators to make suggestions on SOP updates. Take the time to evaluate the suggestions for impacts on safety and quality (in that order) and make the change if neither is an issue. You want to make it easy for operators to succeed. If changes are made to an SOP, make sure they are recorded with reference to what has been updated (see "Organizational details or updates" above), and the changes are clearly communicated and trained on.

SOPs are a fundamental tool for every brewery's quality system. While the sheer number can be hard to manage, the time put into keeping them up to date is well worth it. When implemented, they will support every level of the quality priority pyramid you are trying to build. SOPs should be the first tool in your quality toolbox.

5

CLEANING AND SANITATION

Some of the most important SOPs in your arsenal revolve around cleaning and sanitation. You may hear them referred to as sanitation standard operating procedures (SSOPs), but I will simply refer to SOPs from here out.

Some brewers refer to themselves as yeast janitors; they are not wrong. Not only do cleaning and sanitation procedures prepare the brewery environment to receive wort and yeast, they are also a primary component of your GMP and food safety programs. Understanding what you are trying to clean and how to accomplish that is the first step in writing an effective brewery cleaning SOP.

Cleaning and sanitation are complementary processes. To repeatedly brew good quality beer, these procedures cannot be skipped. When you clean you are trying to remove soils from surfaces, such as the krausen ring, hop particulate, and other types of beer or process residue. When you sanitize, you are reducing the population of microorganisms present, typically through the medium of chemicals, heat, or steam. *Cleaning must always proceed sanitation*. One of my favorite sayings in our brewery is "you can't sanitize dirt." Think about it, microorganisms thrive in a dirty environment, especially one that is nutrient rich such as beer or wort. Cleaning is the first step toward brewing great beer.

Cleaning and sanitation can be accomplished using a bucket and brush, or something more sophisticated such as a clean-in-place (CIP) system. CIP systems are preferred for food contact surfaces, such as brewing vessels. Why CIP? Well, human skin and clothing can harbor microorganisms. These unwanted microorganisms can find their way into your beer and contaminate or spoil it over time. To reduce the risk of cross contamination, the less contact we have with the inside of our tanks the better. To control the shelf life and flavor of your final product, you will want to avoid unwanted bacteria and yeast. These unwanted bacteria and yeast can lead to unexpected flavor attributes, or worse, an exploding can, bottle, or keg. Cleaning and sanitation are your first line of defense against these outcomes.

CLEANING

Before we start to develop our first cleaning SOP, we should discuss some of the factors you must consider when cleaning your brewery.

Soils

First things first, before choosing a chemical or process to clean, think about where the soils are coming from. There are five main sources:

1. Food residue (beer)
2. Water residue
3. Environmental contamination
4. Biofilms, which are protective layers created by populations of microorganisms on a surface
5. Chemical residue, such as that left over from the cleaning process

All these sources are made up of more specific components that can complicate things and influence our choice in cleaning method.

Food residue. Food residue is organic waste matter. It can consist of carbohydrates, lipids, protein, and minerals from the malt, hops, and other ingredients. Food residue is typically deposited onto the walls of your vessels, for example, a krausen ring. Protein and mineral deposits may require a two-step cleaning process.

Water residue. Water residue is typically made up of mineral deposits, or microorganisms if the water is contaminated. If your municipal water supply is safe to drink it is safe to brew with, but be aware how the water is treated before it gets to your brewery. This can be as simple as an online search or a phone call. If you are using well water to brew it is a good idea to have your water tested at least once a year, or more often as necessary, so you are aware of its makeup.

Environmental contamination. Environmental contamination comes from the bacteria, yeast, and mold that live all around us and can unexpectedly end up in our beer. These organisms travel in the air, on our tools, and on our ingredients. They also hitch a ride on us; this is the classic route of cross contamination. We can combat environmental contamination by following good personal hygiene practice, including a "kill step" in our process, and implementing thorough cleaning regimens.

Biofilms. Biofilms form as a result of poor cleaning and sanitation techniques. These films occur when cells in a population of microorganisms stick together and create a plaque or film on a surface. These plaques are difficult to clean and resistant to many normal cleaning methods.

Chemical residues. Chemical residues are left over from the cleaning process. These residues must be rinsed away before proceeding to the next step, otherwise you run the risk of a dangerous chemical reaction or effectively neutralizing the next chemical and rendering it useless. In the worst-case scenario you can adulterate a final product making it unsafe to consume.

All of this sounds intimidating, doesn't it? Well, worry not, every brewer since the dawn of time has had to deal with these same elements.

Knowing the variety of soils, it becomes quickly apparent that not every chemical is created equal. Using the wrong chemical for a certain kind of soil can be like trying to drive a nail with a broken screwdriver. There has been a lot of research and development on the best chemical and process to use for each kind of soil a brewer will face. I would suggest reaching out to a chemical supplier that works with food processing plants. A good supplier should help you figure out how to safely clean up tough messes without using an excessive amount of chemical.

Cleaners

In general, you have five main categories of chemicals that can be employed to clean around your brewery: acid, alkaline, caustic, chlorinated, and detergents. To help you start a discussion with your chemical supplier on choosing the best chemical for the job, here is a general overview.

> ⚠ For your safety and that of your staff and your processing equipment, do *not* mix chemicals or use them for any task for which they are not intended.

Acid cleaners. Acid cleaners have different uses around the brewhouse. They are used for passivation of stainless steel. They also effectively remove mineral deposits like beer stone. Formulated acidic detergents can be effective at solubilizing and removing organic deposits. Acid neutralizes caustic "seeds" and prepares the surface for sanitizers that would otherwise be inactivated by higher pH levels.

Alkaline cleaners. Any cleaner that uses a pH above 7 as a method of cleaning without utilizing caustic soda (sodium or potassium hydroxide) is an alkaline cleaner.

They work by breaking down solids into smaller molecules that become soluble. Although they are not as strong as caustic cleaners, when alkaline cleaners are used appropriately (with care) they are safer on personnel and soft metals (e.g., brass or aluminum), making them useful on draft systems, packaging lines, or other machinery with soft metals.

Caustic cleaners. Caustic cleaners are the workhorses of food processing plants. They have a high pH (≈12.5) and break down solids into smaller molecules that become soluble. Modern caustics are formulated with surfactants and chelating agents to aid in the cleaning process. Different applications may require different concentrations, but a modern caustic formulated for CIP can be highly effective in a 1%–3% solution.

Chlorinated cleaners. Chlorine is an effective cleaner in several ways. It is an excellent oxidizer and a useful additive to caustics and detergents. However, chlorine is falling out of favor with brewers for a few reasons. First, residual chlorine can lead to problems with chlorophenol off-flavors (antiseptic mouthwash) in finished beer. Second, chlorine outgasses from solution at temperatures greater than 150°F. Finally, chlorine cleaners are incompatible with acids and a dangerous gas cloud will be created if the two are mixed.

Detergents. Detergents, such as store-bought dish soap or powdered bathroom cleaner, are not regulated for food processing. While they are generally effective, detergents should not be used on processing equipment. They can also be quite costly when used on a large scale. Detergents can be reserved for small-scale dirt removal, but you would be better off looking into an alkaline cleaner for the same application.

SANITATION

Sanitizers do not clean, but rather work to kill any microorganisms left behind after a proper and thorough cleaning. A good no-rinse sanitizer will probably be the most used chemical in your brewery. Sanitizing is defined as reducing microorganisms to a safe level on clean surfaces. Choosing the right no-rinse sanitizer is important. To comply with FSMA and your GMP standards, your sanitizer should be documented to perform a 5 log reduction in microbial load. That is equal to a 99.999% reduction. Luckily, you do not have to prove this yourself, all the information you need should be on the safety data sheet (SDS) provided by your chemical supplier. First check the top of the SDS to see if the product is listed as a no-rinse sanitizer appropriate for your industry. Next, check the SDS for an Environmental Protection Agency (EPA) registration number under the name of the product.

That was easy. Now you must choose the right sanitizer for the job and make sure your working solution is in the appropriate parts per million (ppm) range using test strips or titration. The no-rinse application is an important distinction—a more concentrated sanitizer is not always better. If you are above the working range of the no-rinse sanitizer then it should be rinsed off with water. Keep in mind that rinse water could also introduce microorganisms to the inside of the tank, which would defeat the purpose.

There are three forms of chemical sanitizer and one physical sanitizing method that I will mention:

1. Oxidizers
2. Acid anionic
3. Surfactant
4. Steam

Oxidizers. Possibly the most popular sanitizer type among brewers, oxidizers "burn through" cell walls, denature proteins, break sulfur bonds, and inhibit DNA replication. As with every chemical, consult the SDS and use only as intended (table 5.1). These chemicals can be as harmful to you as they are to the microorganisms you are trying to control.

TABLE 5.1 COMMON TYPES OF OXIDIZING SANITIZER

	Oxidizing sanitizer type			
	Peroxyacetic acid	**Chlorine**	**Chlorine dioxide**	**Ozone**
Applications	CIP, spray, and waste-water	Broad spectrum	CIP, COP, and spray	Applied directly to water and surfaces
Advantages	Single chemical addition; wide range of efficacy; direct food contact allowed; eco-friendly, breaking down to oxygen, carbon, and acetic acid	Inexpensive and easy to apply	Has mechanism to degrade biofilms; direct food contact allowed. Dilute solutions can be stored for up to 72 hours before they lose efficacy. Oxidizing agent or chemical species is chlorite, not chlorine	Dissipates quickly; eco-friendly
Disadvantages	Vapor pressure issue makes it dangerous to breath and handle; harsh on skin and other tissue; incompatible with soft metals; potential beer oxidation issues	Potential chlorophenol off-flavors; harsh on environment; hard on equipment; loses efficacy quickly in dilute solution	Mixing without proper equipment can create harmful gas; mixing equipment can be expensive	Expensive; very corrosive, reactive, and unstable. Each machine must be registered, and registration does not mean it is certified by EPA

CIP, clean-in-place; COP, clean-out-of-place

Acid anionic. Acid anionic sanitizers disrupt cell membranes and denature proteins. The classic example of acid anionic sanitizers are iodophors (iodine + acid), which are good for CIP, clean-out-of-place (COP), and hand sanitizing dips. Iodophors provide a broad-spectrum kill and are the least harmful to skin. The pH of the sanitizer solution should be below 3. The acid involved allows the product to be tolerant of hard water and its brownish yellow color makes it easy to identify in solution. Be aware that iodophors can contribute unwanted chlorophenol flavor attributes and can stain equipment.

Surfactants. Due to their strong positive charge, surfactant-based sanitizers disrupt the cell membranes of microorganisms, causing them to rupture and die. Quaternary ammonium sanitizers are the classic example used in breweries. They are great for tank exteriors, walls, and floors. They foam well and are great for surfaces and are safe on soft metals. However, the foaming capabilities present a serious issue in CIP applications. Residual quaternary ammonium on surfaces can be a food contact issue.

Steam. Steam can be used as a sanitizer in your brewery. Steam is high energy and will kill microorganisms on contact. In an autoclave, steam is used in conjunction with pressure to sterilize equipment and is great for small-scale sanitizing/sterilizing in the lab. Steam can be expensive to produce and hard to control on a larger scale but it can be done. However, steam poses a serious scald risk to staff, can melt soft parts, and cause pressure/vacuuming issues in the brewhouse.

APPLICATION VARIABLES

Regardless of the cleaning and sanitation method (CIP, COP, foam, or scrub), there are five main variables to consider and control:

1. Water
2. Temperature
3. Chemical concentration
4. Mechanical action
5. Time

Balancing and controlling these variables will lead to a successful and efficient clean. Let us look at these variables in detail.

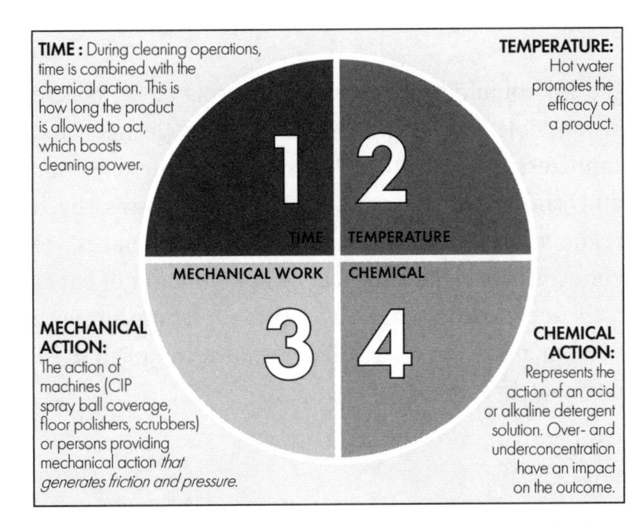

Figure 5.1. The Sinner's Circle represents major variables responsible for the success of a cleaning process.

Water

Water is the solvent for your cleaner and for the soil. It is also the medium for pre-clean and post-clean rinses. The water used must be potable. Know your water source, because the salts and metals contained in hard water can react with the cleaning chemical, causing mineral deposits and staining.[1]

Temperature

Increasing the temperature of your cleaning solution will increase the efficacy of water as a solvent and increase the rate of reaction of soil with your cleaning chemical. Higher temperature is not always a good thing though; recent research suggests with modern dry hopping rates higher temperatures can "bake on" residual hop oils.[2] Also make sure the chemical you are using is designed for high temperatures and will not outgas, rendering your cleaning/sanitation process ineffective.

Chemical Concentration

Starting with the right amount of chemical is imperative. As straightforward as it sounds, adding a measured amount of chemical to a volume of liquid is not always foolproof. You can confirm the concentration through titration. A titration is simply adding a known concentration of chemical to an unknown solution. The results will allow you to calculate the concentration of chemical in your cleaning solution. In this application, a drop titration kit is usually appropriate. (Titration will be covered in detail in chapter 19.) A good chemical supplier can supply you with a titration kit for just about any chemical you are working with; alternatively, you can go use third-party companies for titration kits. Be aware that the concentration of your cleaning chemical will decrease as it reacts with soils or other constituents (such as CO_2), so check your chemical concentrations before you start the clean, especially if using equipment that reuses chemicals, for example, CIP skids or keg washers.

Mechanical Action

Mechanical action requires adequate physical contact and action with the surface to be cleaned. This is easy to control when using a bucket with a scrub brush. Inside a closed vessel that is part of a CIP system it is more difficult. Some manufactures will provide tank, spray ball, and pump specifications. This is to ensure turbulent flow, which is achieved at an optimum flow velocity through the system. The gold standard is 6.5 ft./s (2 m/s). Flow velocity less than this may not provide sufficient mechanical action, and above this there is no additional cleaning benefit. Making sure your equipment can meet flow velocity requirements will allow you to achieve good mechanical action on all surfaces of your vessel. You can calculate flow velocity with the following equation:[3]

$$v = \frac{4 \times Q}{d^2 \times \pi \times 60}$$

where

v = velocity in feet per second

Q = flow rate in cubic feet per min.

d = pipe diameter in feet

1 Michael Lewis and Tom Young, *Brewing*, 2nd ed. (New York: Kluwer Academic/Plenum Publishers, 2002), 139.

2 "Hop Oil removal Cleaning Procedures," Blog, Alpha Chemical (website),November 14, 2018, https://alphachemical.com/blog/hop-oil-removal/.

3 Rench, Richard J. 2019. *Brewery Cleaning: Equipment, Procedures, and Troubleshooting*. St. Paul, MN: Master Brewers Association of the Americas.

Time

Appropriate contact time with cleaners and sanitizers is integral to their efficacy. Each chemical has its own recommendations. Work with your supplier and check the SDS to find out these contact time recommendations. If you stick to these guidelines as a minimum contact time, you will be in good shape.

The overall goal of cleaning and sanitation is to prepare the equipment used for your process by reducing the microbiological load to zero or as close to zero as possible. There are no single-step cleaners and sanitizers available to do this. To reiterate, cleaning and sanitation are complementary processes. Each regimen has its advantages and disadvantages, so choosing the right chemical or combination of chemicals and building a good working relationship with your chemical supplier will help you accomplish these goals safely and effectively.

ADENOSINE TRIPHOSPHATE (ATP) MONITORING

Adenosine triphosphate (ATP) luminometers are devices used in conjunction with ATP swabs for monitoring plant sanitation. Luminometers use bioluminescence to detect residual ATP as an indicator of surface cleanliness. ATP is the universal unit of energy used in all living cells. The presence of ATP on a surface indicates improper cleaning, including beer residue, allergens, and/or bacteria.* ATP monitoring is easy to do and relatively inexpensive. In conjunction with your microbiology program, it can be a valuable tool for validating your CIP procedures.

HOW IT WORKS

You collect a sample using a surface swab or honeycomb-shaped water collection tip (depending on which test you run). I suggest using surface swabs while getting used to this new technology. After collecting the sample, carefully return the swab to the included test tube and add the reagent as directed in the instructions. The reagent containing luciferin/luciferase will react with the ATP to generate quantifiable bioluminescence that is detected in the luminometer. Low relative light units (RLU) indicate a clean surface while high RLU indicate contamination.

SANITATION LOGGING AND SETTING LIMITS

Each brand of luminometer uses a different scale for RLU, so commissioning a unit is imperative to set acceptable limits in your brewery. Test both clean and dirty surfaces to get an idea of the scale you are working with. For example, a meticulously cleaned stainless steel surface should be below 10 RLU, a harder to clean surface may be around 25 RLU, and a dirty floor with standing water will be around 2,500 RLU. Use these tests to set limits based on your manufacturer's recommendations because these RLU ranges can vary greatly between manufacturers. Include ATP testing as a routine validation of your cleaning procedures. This can be a helpful QCP to ensure time and chemicals are not being wasted on an ineffective CIP and dirty tanks do not get filled.

* "Frequently Asked Questions," Hygenia (website), accessed 2018, s.v. "What is ATP Monitoring?" https://www.hygiena.com/frequent-asked-questions-food-and-beverage.html.

PART II
BUILDING YOUR QUALITY SYSTEM

As I stated in chapter 2, QA/QC is not a stand-alone program but part of a larger quality system. We saw in the quality priority pyramid (fig. 2.1) that GMP and food safety make up the foundation of your system, and for good reason: if you are not producing a product that is safe and fit for human consumption it does not matter if you have tight process controls or standards. So, with your GMP manual and food safety plan as your foundation, it is time to start building the framework to support your quality system.

There are two major components to your quality system, the quality manual and the GMP and food safety manuals. We started with GMP in part one because it is mandatory and likely more unfamiliar to many brewers. Equally as important to a robust quality system is the quality manual. The quality manual is where we start to put together a quality plan and lay out instructions for its execution. Your quality manual will work in tandem with your GMP manual, both will act as instructions for the execution of your quality system.

Let us start to look at the full view of your quality system. This includes all the policies, manuals, tasks, and measurements you perform and use to make quality beer. The quality manual, which is what we will look at in the following chapter, is where you will lay out your quality plan by identifying process controls and setting standards that are important to your brewery. Just like your GMP and food safety programs, you will use SOPs to execute your newly written quality manual.

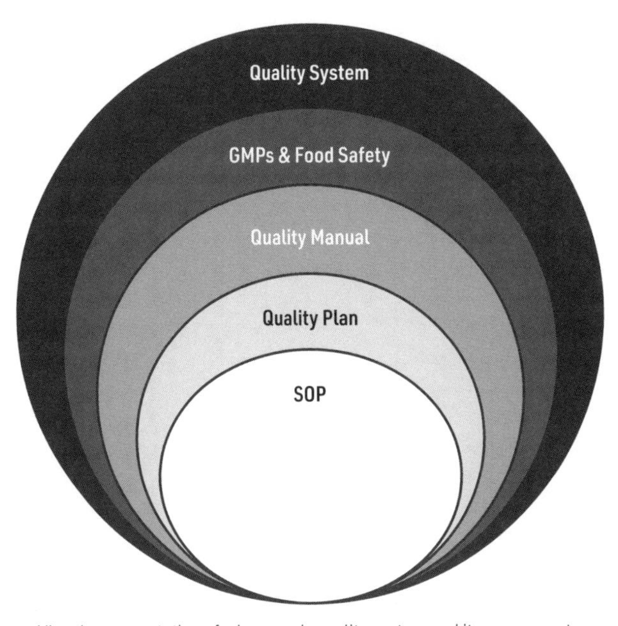

Visual representation of a brewery's quality system and its components.

6

STARTING YOUR QUALITY PLAN

In part 1 we focused on GMP and food safety, the first two tiers of the quality priority pyramid (fig. 2.1). The next two tiers are process controls and standards, both of which require measurements of some kind to be made. While it is easy enough to get a few measuring instruments and obtain data, that data will not be useful without some planning and purpose. This is where your quality plan comes into play—you plan what measurements to take and where to take them. In other words, you identify process controls in your quality plan. Your quality manual will provide the guidance needed in order to execute your quality plan. This chapter will focus on these two integral parts of your quality system. In preparation for writing your quality manual, you will start with planning what process controls you want to measure as part of your quality plan. Process controls are the parts of the brewing process that you measure regularly and consist of various product parameters that can be controlled throughout the brewing process, from raw materials to packaged beer to product handling outside the brewery.[1] When you measure process controls you can make adjustments—either on the fly or for the next batch—to positively influence the quality and consistency of your beer. From here on out, I will refer to process controls as quality control points (QCPs). Sound familiar? These are critical control points (CCPs) for your quality manual.

COMING UP WITH YOUR QUALITY PLAN

Throughout the brewery you will use in-process measurements to make decisions that directly affect the outcome of your product. Planning what process controls are the most important to monitor can be a difficult task, especially if you are like me and want to know everything about your beer all the time. To me, this is why a quality plan is so important: it helps you focus on what is important to your process. Not every process control or QCP holds the same importance for every brewery, but there are some non-negotiable measurements that every brewery must make. I will do my best to present them in part 2 of this book, then cover the methods and tools to actually make these measurements in part 3. Beyond that, I will also provide you with some insight on what to choose for your brewery.

Let us get right to it and point out the non-negotiable measurements. If your brewery is brewing beer to sell to the public, you should be able to measure:

---

1 "Quality Priority Pyramid," Educational Publications, Brewers Association (website), accessed July, 2018, https://www.brewersassociation.org/educational-publications/quality-priority-pyramid/.

- Weight of grain, hops, and many other constituents of your beer. The scale will range from many kilograms or pounds to just a fraction of a gram.
- Volume of liquids
- Density/specific gravity
- pH
- Yeast viability and slurry density (if re-pitching)
- Sensory analysis
- Concentration of cleaning chemicals

In addition to the above list, if you are a packaging beer—including mobile canning/bottling lines—and it leaves the walls of your brewery, you should be able to measure:

- Can seam and bottle crown seals
- Carbonation
- Dissolved oxygen (DO)
- Presence of spoiling microorganisms, e.g., *Lactobacillus*, *Pediococcus* (part of the microbiology program)

You may be thinking, "Dissolved oxygen meters are so expensive!" My response to that is, "So are canning and bottling lines." Look at this instrument as an investment in quality, and make sure your mobile canning company is using one too. You have spent two-thirds of your brewing process doing everything in your power to make the best beer possible. Why leave the last third (packaging) to chance? Budget a dissolved oxygen meter into the cost of your packaging equipment. The shelf life of your beer depends on it.

Budget a dissolved oxygen meter into the cost of your packaging equipment. The shelf life of your beer depends on it.

When it comes to choosing measuring instrumentation or tools, a good rule of thumb is to budget a measuring tool into the purchase of each piece of equipment. Think of your new equipment as a fast car. If you did not have a speedometer to tell you how fast you were going, what is the point? If you are investing in a new piece of equipment you should be able to measure if that equipment is working as advertised. Having the ability to measure equipment performance is the only way to truly tell if the equipment is working correctly and it will help when planning for upgrades. You may also choose or need to upgrade instrumentation to achieve better accuracy, precision, or for ease of use.

Quality Guidelines

We can now start some serious planning. When you start planning for quality it helps to pre-set some rules for yourself. These rules can be somewhat philosophical but should reflect your overall goals for your final product. I call these philosophical rules your quality guidelines. These guidelines should be used to influence your decision-making when writing your quality plan and its associated manual, as well as making quality-based decisions every day in the brewery. These will be different for each brewery. However, I consider guideline number one to the bedrock of every quality plan: ***safety***.

All breweries are producing a product that is intended for human consumption. The overall safety of your consumers and your staff is paramount when making any decision regarding your processes and product. You need to make decisions that result in world-class beer that is free of physical and chemical adulteration. I discussed how to implement a food safety plan in chapter 4 for craft brewers. Now get to it! Objectively evaluate your manufacturing processes and follow the templates and you will be in good shape.

The order of your next quality guidelines is up to you and can differ from brewery to brewery. Writing down and referencing a few simple quality guidelines will help build the framework in which you and your team make decisions and view quality in your brewery. For example:

Brewery X Quality Guidelines
Safety
Provide a quality experience
Achieve consistency

If this example does not align with your brewery, ask yourself what your big-picture brewing goals are. Find your core values and use those to inform your decision-making. The quality guidelines above are the same that I follow to this day. Inevitably, things will go wrong or you simply stray away from your original plan. In those cases hard decisions have to be made, which is when I look to my quality guidelines to decide the best course of action.

For example, there was a day where one of our staff did not follow our glass policy (i.e., no glass in the production area). After an accident, broken glass ended up in the brew kettle. I went straight to my quality guidelines to determine the best course of action to take. All I had to do was look at the first guideline: safety of the product and the staff. Since we could not guarantee without a doubt that glass would not end up in the final product, it was decided to dump the batch, rigorously clean the kettle, and start over.

Another, less dismal, example would be when a new employee dry hopped our core lager following our IPA dry hop recipe (the lager recipe did not originally contain dry hops). This was caught during the next day's round of gravity readings. Going back to the quality guidelines the product was still safe, so we continued to let it finish on hops. When finished, we tasted it as a group and found the resulting beer to be pleasant. While the sensory analysis revealed it was not consistent with true-to-brand descriptions, that is, it did not strictly meet the *achieve consistency* guideline, it was determined that the resulting beer *provided a quality experience.* So, using our quality guidelines to inform us, we made the decision to keg it as a limited release and the batch was saved. Not only that, the expectations of our consumers who liked the lager recipe as it was were not affected due to an unexpected (although temporary) change in the core brand.

IDENTIFYING QUALITY CONTROL POINTS

Now you have an idea of what non-negotiable process measurements should be taken, and armed with a few guidelines for decision-making, we can start identifying QCPs. You can start by following your process from grain to glass. List all points where a qualitative or quantitative measurement can be taken (see sidebar). Narrow it down to those that directly affect your final product. It is important to take this first step so that your quality plan can be built on, growing with you as you become more experienced or as processes become more complicated. With the understanding of some basic measurements you can take your quality program a long way.

QUALITATIVE VERSUS QUANTITATIVE

A quantitative measurement produces a numerical result, for example, measuring that a beer is 7 SRM. Results are usually obtained by using measuring instruments of some kind, a spectrophotometer in the case of an SRM reading. Quantitative measurements are very reliable if the number is achieved using proven and validated methods, and are easy to chart or graph.

A qualitative measurement describes an attribute belonging to the sample and is not expressed as a number. It usually refers to human sensory experience. For example, the beer in the glass is golden yellow in color. Qualitative measurements can be subjective but can still be very powerful, especially when it comes to building true-to-brand sensory descriptions and staff training.

The example on the next page follows a small packaging brewery's quality control plan from grain to glass with suggested QCP or two.

Quality plan step	Suggested QCP
Evaluate process water daily, checking all sources	pH for chemical contamination Visual assessment for particulates Sensory for off-odors and flavors
Evaluate all raw ingredients	Visual assessment for pests or other inclusions Sensory assessment for condition and freshness
Keep detailed brew logs	Record all brewhouse, cellar, and packaging measurements
Check yeast viability and slurry density for every brew	Cell density count before pitching Cell viability assessment before pitching
Daily measurements of beer in fermentors	Density (specific gravity readings) pH Temperature
Forced ferments on all new brands or recipe changes	Force fermentation test
Microbiological screening	In-fermentor screen for know beer spoilers. Hsu's *Lactobacillus* and *Pediococcus* medium is a great place to start
Check carbonation on every batch before packaging	Volume meter reading in brite tank and before packaging Qualitative sensory assessment
Taste beer prior to packaging	Go/no-go sensory assessment with trained staff
Regular seam or crown checks	Ensure physical dimensions are withing manufacturers specifications using crown gauge, calipers or micrometer
Package dissolved oxygen	After starting up and periodically throughout a run
Record package carbonation, density, pH, and yeast cell density	Volume meter for carbonation Density with hydrometer pH meter Cell density using microscope
Date and lot code all packaged beer	Use unique batch and date identifiers Get packages dated by hand or automated methods
Keep "library" of packaged beer for at least the duration of its shelf life	Try to simulate market conditions in storage If you are limited on space (like me), choose a representative sample of each run that works for you. I started by keeping one 4-pack from each canning run and elected not to retain any kegged product, though I wished I could
Use third-party labs to confirm internal lab results or perform tests out of your lab's reach	Send samples out for analysis (microbiological, chemical) if problems are detected or there are gaps in testing procedures
Verification/calibration of thermometers and other instruments on a regular schedule	Calibration of equipment as often as needed based on manufacturer's recommendation or measurement's importance
Implement a sensory program	Schedule regular tastings, focus on true-to-brand Work to include attribute training

Let us dig a little deeper. To start your written quality plan, take each quality plan step you choose and state the objective of the measurement or check. Assign and communicate who is responsible for the measurement or check, being as specific as needed. Be explicit on how to execute the check. By including an SOP you spell out exactly how to measure the associated QCP. Each quality plan step should be important enough to be recorded, so be sure to specify where it is recorded. For example, let us look at our first quality plan item, the daily evaluation of process water:

Evaluate process water daily	
Objective:	Ensure that process water meets requirements for brewing and cleaning. Water must be free of sediment, discoloration, odors, and flavors that are not native to your water profile.
Responsibility:	First-in brewer
Execution:	Via pH check and sensory assessment. See Water Quality SOP.
Tracking:	Go/no-go water quality log located on HLT.

Each measurement or check should have an objective like the one above. It is also important to remember that if you are taking the time to make this quality check, you should train the staff on it. Regarding training on water evaluation, every region will have its own profile. Keep it simple and stick to two or three of the most common issues that can occur in your local water supply. In my brewery, sediment from our city water system is seen often enough that we train on that by emphasizing visual sensory for our water quality check. Other common sensory attributes for water are things like hardness, chlorine levels, and possibly sulfur.

Let us look at one more example, this time for daily measurements on beer in fermentors:

Daily measurements on beer in fermentors	
Objective:	Ensure fermentation is proceeding as expected by monitoring gravity, pH, and temperature of every beer in process, every day or more frequently if necessary.
Responsibility:	Cellar crew
Execution:	Cellar crew will take a sanitary sample and record the results using Gravities with the Hydrometer SOP.
Tracking:	Fermentation log located in "Beer in Process" binder and archived in production tracking binders.

SAMPLING PLAN

Now that you have a quality plan outline, you can take a closer look at the execution part of this plan. This revolves around a purposeful sampling plan. A sampling plan is a detailed outline of which measurements will be taken at what times and by whom. A sampling plan describes the direct application of your quality control plan and includes in-process measurements.

With your quality control plan in hand, start to identify quality control points (QCPs) that relate to your quality plan steps. This is the same process we used when identifying hazards in our food safety plan. This time our motives are different—we are looking for measurement points (QCPs) that control the final outcome in some way. For example, the quality plan above lists the step "Check carbonation on every batch before packaging" and one of the associated QCPs is the carbonation volume reading in the brite tank. This QCP affects your quality goals: if the beer is undercarbonated it will negatively affect consumers' sensory perception of the beer; if it is overcarbonated it will not pour on the majority of draft systems. Therefore, you must set a standard, in this case, the standard is 2.5 volumes CO_2. This QCP meets all the audit standards. Using my instrumentation (volumeter) I can react to the reading by adjusting my carbonation based on my results. In later chapters, I will point out possible QCPs for each measurement discussed in part 3. Keep in mind not every QCP outlined there will be relevant to your process. Choose QCPs that are relevant to your quality guidelines and that you can react to. To help you decide, audit a QCP by answering a few questions about the specific control point. If the answers are positive and make sense to you go ahead and start measuring.

Sampling plan QCP audit
What is the impact of this measurement to my quality goals?
Do I have an accurate and reproduceable way to measure this?
Do I have the ability to react to this?

Now that your QCP is identified, the staff needs to be trained on it. This is where you write an SOP on how to take the measurement. Also, and this is an important distinction, make sure you reference the measurement SOP within the overall process SOP, like the example of the "Gravities with the Hydrometer" SOP in the execution step of the QCP objective above for measuring beer in fermentors. This makes it easy for the operator to look up each SOP they need to accomplish their task. Make sure to include important points about the measurement, for example, is it time or temperature sensitive? If so, make sure all operators know *when* to take the measurement. SOPs should be developed on how to take measurements for all processes in your quality control program. Uniformity through well-written SOPs is key to a successful sampling plan.

The final step to put your sampling plan in action is to include a field for the measurement to be recorded on your brewing and process documentation. The information will only be useful if it is recorded. Also, think about recording variables that influence the QCP you are recording. For example, if you are recording gravity at the start of boil, measure how much water is added to the mash at each step. You can adjust this parameter to influence your boil gravity (fig. 6.1). Group measurements in a logical order on process records. Keeping these measurements in order will help with the flow of the process and mitigate the chances that someone will forget it.

PULLING IT ALL TOGETHER

You now have the tools to build the framework of your quality system. Keep it simple and start by outlining your current measurements and procedures and then add to it from there—this is a continuously evolving program. As you get more experience, and acquire more staff and resources, add new points to your quality program using the methods outlined above. Your quality plan is a living document and should be updated often. Just as a good brewer is always trying to brew the next batch better than the last, a good quality scientist is always improving their measurements and processes.

Brand			**Unique ID#**			**Brew Date**			**Fermentor**		
IPA			1821			3/4/2019			FV 2		
Grain Bill											
Lbs.	Ingredients					Lot #					Initials
800	Two Row Pale					MBL-2674					MW
220	White Wheat Malt					201-0075					MW
50	Carapils					RO 1201819					MW
Mash Targets		Grist Lbs.	Total gal	Strike gal	Flow rate (H2O)	Time flow	Strike temp	Target Mash Temperature			
		1070	680	315	4.1	21	162	150			
Mash Data		Mash in time	Mash end	Sach rest end	Base water	Strike gal	Strike temp	Mash pH	Mash temp		

QCPs

Figure 6.1. Example brew log sheet with potential QCPs highlighted.

7

WRITING A QUALITY MANUAL

After six chapters of thinking about and writing policy and putting the pieces of your quality system in place, this is where it all comes together. Your quality manual will set the guidelines for executing your quality control and assurance program to support the quality system. Your quality manual will also guide you through tough decisions, like what needs to be done when a beer is out of specification. Even more important is who oversees making those decisions and who roots out the cause of the problem.

The simplest version of your quality manual will contain a description of your brewery's values in terms of quality, and a description of roles and responsibilities as they relate to quality governance and management. It will contain the specifications of finished products and when they are to be reviewed, as well as a written plan for dealing with a product that is out of specification. The quality manual will include or reference your GMP and food safety programs. Remember, your quality manual is a living document—as pieces of the quality system come together you should add them to your quality manual as appropriate.

Quality Manual Table of Contents
Quality policy statement
Quality governance and management
Standard procedures GMP Food safety Master sanitation schedule Quality program
Product Specifications In-process targets True-to-brand description
Consumer and In-House Complaints
Corrective Action Plan
Good Laboratory Practices
Calibration Plan

QUALITY POLICY

Until this point, we have been laying the foundation and setting out the framework for your quality system. Your quality manual is where the bricks are laid so that the structure takes shape. You can think about housekeeping duties as executing your quality system. According to Mary Pellettieri, "Quality policy, for our purposes, is the way to behave and to honor and execute the quality system."[1] I could not agree more with this statement. If we want to keep our nicely decorated home neat, we need to write down and implement quality policy. This will become more important as the brewer goes through tough times, changes management, or growth happens. By writing down a brief policy, management, whoever it is, can quickly refer to it and make sure you are staying true to intentions. Your quality policy , which will be written in your quality manual, is the beacon you follow to find your way back home. Ultimately, your quality policy is totally up to you. It only has to be a few sentences, so answering a few questions for yourself may help realize your own quality policy. Here is an example of how we determined our quality policy statement when I was at Rising Tide Brewing Company:

Q: What is important to you and your brewery in terms of quality?
A: Producing the best beer possible and providing a safe and optimal experience in each glass.
Q: What is the ultimate goal?
A: Consistently producing world-class beer.
Q: How do you expect to get there?
A: Determining and observing GMPs and best practices; and dedication to continual improvement.

And here is an example of a quality policy statement using the answers to these questions:

Quality Policy

At Rising Tide Brewing Company, we strive to make world-class beer. In order to do so, our entire team, brewhouse, cellar, packaging, operations, and quality lab, must observe GMPs and best practices for all tasks. We want to provide an optimal experience in each glass of beer and are committed to continuously improving our processes and systems to achieve a world-class reputation.

Your quality policy should align with your company values. What are your company values? Do you value education internally or education of the public? Social justice? Brewing eclectic or innovative styles or using unique ingredients? Do you like mastering classic styles? These values may not directly address the quality of the beer you are producing, but they can help inform your quality policy as you are writing it.

[1] *Quality Management: Essential Planning for Breweries* (Boulder: Brewers Publications, 2015), 18.

QUALITY GOVERNANCE AND MANAGEMENT: ROLES IN IMPLEMENTATION

Two major goals in writing a quality manual are assigning management roles and communicating them to all other staff. When it comes to assigning roles there are two specific kinds we are interested in, governing roles and managing roles. **Governing roles** are typically the high-level decision-making type including but not limited to:

- setting specifications for the brewery and its brands;
- establishing measurement procedures, such as where, when, and how;
- setting policies (e.g., final say in role assignment) and procedures (e.g., SOPs) and changing these when necessary;
- setting goals for quality improvements.

Managing roles are the daily decisions assigned to department heads or key employees, including but not limited to:

- the act of measuring and reporting;
- implementing and documenting training;
- managing daily operations;
- implementing corrective action on out-of-specification or control products;
- data review and continuous improvement.

There may be a lot of overlap of these roles in a small craft brewery, which is perfectly OK. The important part is that management and employees are aware of their roles and of who makes decisions when things go wrong, because things will go wrong. One of the biggest advantages of having your quality system in place is the ability to flag and correct these issues before your beer makes it out the door. Once the beer is out the door you have no control over it and a small oversight can turn into a lot of lost revenue and consumer/retailer trust.

After thinking about upper-level management's roles, let us take a closer look at the roles brewhouse, cellar, packaging, and quality teams play in your QA/QC system. Set out guidelines that staff are expected to follow. This is not the place to get nitpicky or micromanage, but to instead point out general expectations.

Department Responsibilities

Brewhouse team: Responsible for carrying out wort production safely and according to SOPs and quality system principles (GMP, food safety, and quality programs).

Cellar team: Responsible for supporting the brewhouse, preparing fermentors for receiving wort, monitoring fermentation, harvesting yeast in conjunction with the lab team, and conditioning and filtering. All activities should be performed according to SOPs and quality system principles.

Packaging team: Responsible for preparing brite tanks for receiving green beer; timely carbonation and packaging of beer in kegs, cans, and bottles; and preparing beer for shipment to warehouse or distributor. All activities should be performed according to SOPs and quality system principles.

Quality team: Responsible for the implementation and maintenance of quality system principles. Implement, train, and oversee QCP measurements and calibrations. Maintain and execute a sensory program. Maintain and execute a microbiology program.

Two important points here. First, for this to work, SOPs and quality systems should be in place or in the works. This is a fantastic opportunity to work with your team on creating SOPs. Often those who perform a task every day can articulate it with ease. Managers can review the SOPs through the lens of GMP and food safety and make sure the process is sound before training others on the official SOP. Second, there may be a lot of overlap of these tasks and teams. At Rising Tide, all brewers are cellar operators and have limited training on the canning line. Our packaging operators are also cross-trained in cellar operations. Clearly defining these teams (brewhouse, cellar, and packaging) and their associated daily responsibilities will prepare you for growth and prevent arguments. It can also help prevent important tasks from falling through the cracks if everyone is clear about their role for the day. Other departments can be added or swapped in to reflect your operation as it stands or grows. Clear communication, or lack thereof, can make or break your day.

STANDARD PROCEDURES

Standard procedures describe how you are implementing your quality system. They include everything we have been working on up to this point: your GMP program, food safety program, master sanitation schedule (MSS, more on that later), and quality plan fit here. These will make up the bulk of your quality manual. Include all these plans and assign responsibility for the execution of each plan. Depending on the size of the brewery, responsibility can be spread out to different department heads. If an SOP exists for the task, reference it. If you are performing a task, it should be recorded somewhere. Point to where that record lives. Be transparent and direct. Your employees will be appreciative of the clarity and so will any potential auditor.

Master Sanitation Schedule

Before you even started to think about your quality manual, you knew that cleaning was of the upmost importance. You were right. As you grow and get busier, you hire more help. Do your coworkers know how important cleaning is to make quality beer? Formalizing procedures for cleaning in all departments is the first step in highlighting the importance of cleanliness. The SOP gets us part of the way there, but the master sanitation schedule (MSS) will explicitly address what is to be cleaned when and by whom, as well as what to do if a cleaning regimen is not successful.

When starting to write your MSS, prioritize procedures that have direct contact with beer preparation. The first thing that comes to my mind is CIP of all brewing vessels. When it comes to something this important, it is useful to explicitly define the expectations of the operator performing the CIP. This can be done in a paragraph such as this:

CIP Responsibility

The cellar, production, or brewhouse staff conducts all CIP procedures according to SOPs. All production staff, in collaboration with the quality team, performs process verifications of chemical levels, rinse effectiveness, and cleaning validation. In the case of a failed CIP cleaning, as shown by a verification procedure (pH, swab or visual), the immediate corrective action in the form of a re-clean, rinse, or other is the responsibility of the production crew representative. All CIP documentation is the responsibility of the operator performing CIP.

This leaves no question on what must be done even when the CIP is not successful. Once the operator's responsibilities are laid out, it is time to lay out the actual policy surrounding the cleaning of each vessel. Not every cleaning procedure is the same for every vessel; instead, this is the place to give a general outline of the cleaning procedure and point to the appropriate SOP for exact details. Note that the examples below show the relevant SOPs underlined. Whether digital or paper, SOPs should be uniquely named and/or numbered, including the version. Correctly and clearly reference or link to the most up-to-date version so that there is no confusion over which SOP staff should use for a given task (see chap. 4, p. 37).

Sanitation/CIP Policy

The brewery is to be kept in good working order under strict cleanliness rules as dictated by our GMP and MSS policies. CIP processes are routinely conducted with the following requirements:

Brewhouse vessels. Brewhouse vessels are cleaned with caustic at the end of the brew day according to the Brewhouse Cleaning SOP. At the end of the brew week an acid/caustic cycle is conducted according to the Brewhouse Cleaning SOP.

The grist mill is cleaned weekly or as determined by the head brewer using the Grist Mill Cleaning SOP.

Cellar vessels. Fermentation vessels are cleaned using a caustic wash after emptying, then sanitized with a peracetic acid cycle prior to filling. Cleaning is conducted according to the Fermentor/Unitank Cleaning SOP. Removal and cleaning of all valves, gaskets, etc. is part of the CIP process. A pH reading is taken of the rinse water to ensure no residual caustic remains in the tank. In addition, an inspection, both visual and with ATP swabs, is required with every CIP. The results are recorded in the appropriate sanitation log and filed in our SOP binder under the **At Source Records** tab and logged digitally. Before moving wort or beer into a fermentation vessel the vessel must be sanitized according to the Fermentor/Unitank sanitizing SOP. Tanks left empty for more than 24 hours need to be re-sanitized.

Brite beer tanks. Brite beer tank cleaning, rinsing, and sanitizing methods vary depending on the next beer in line to be packaged, as well as the length of time the tank has been left empty. The specific tank handling method is laid out in the current Brite Beer Tank Cleaning SOP and the Brite Beer Tank Sanitizing SOP.

Packaging equipment. The canning line is sanitized with peracetic acid before operation and cleaned with an alkaline detergent solution immediately after the brite beer tank is emptied according to our canning line SOP.

Kegs are cleaned and sanitized prior to filling using the Premier three-station semi-automatic keg washer. The keg manifold is sanitized with peracetic acid prior to filling and immediately cleaned with an alkaline detergent solution after use according to the keg filling SOP.

The next item to address is documentation of CIPs. This documentation is important. These records communicate to the next operator that the tank is ready for the next step and they will show an inspector that you are following your own protocol and working toward complying with GMP.

Documentation

All CIP cleanings are documented on a sanitation log in the "pink binder" for tanks, with the canning line, and with the brewhouse documents. When the sanitation log is full, it will be placed in the SOP binder under the **At Source Records** tab and logged digitally. Chemical concentrations will be tested by the quality manager or appointed representative using a titration kit or similar every month. ATP swabbing in conjunction with an in-house microbiology sampling plan will verify effectiveness of the CIP procedure. Any high counts or areas of concern are investigated, and corrective action is taken and documented by the quality control team. Microbiological documentation can be found electronically, or in the lab results binder.

Keeping these records, you can easily spot trends and adjust your procedures to accommodate hot spots, which will eventually lead to time and money saved.

Having a maintenance plan in place for CIP equipment is especially critical. You cannot clean effectively with a broken mop, and the same goes for a damaged or clogged spray ball or broken pump. Schedule maintenance and inspections for CIP equipment regularly. Make sure there is a specific person in charge of ensuring this is done.

Maintenance of CIP Equipment

CIP equipment is maintained as process equipment by the cellar and maintenance team. Inspection and preventative maintenance plans are the responsibility of the maintenance team and head brewer.

This discussion so far covers some critical CIP areas directly relating to QA/QC. For the rest of the MSS, prioritize high-risk cross contamination areas, such as floors, drains, and work surfaces, as well as GMP tasks. The details do not have to be as involved, but you want to make sure there is a person appointed, state how often it is to be done, and give the procedure/chemical to be used.

Floors. Floors are to be kept clean throughout each daily task. Spills will be cleaned immediately. Work areas, tile, and grout is cleaned with a scrub brush and appropriate chemical, and rinsed with water at the end of the task. At the end of the brew week all floors are cleaned with industrial floor cleaner and then rinsed with water.

Your MSS is just as critical as your GMP and food safety programs. Getting it written and implemented will improve the quality of your beer and reduce the risk of cross contamination, ensuring that your brewery is ready for presentation tours and the inspectors. For tours, if your facility looks clean, the consumer will associate that with quality beer.

PRODUCT SPECIFICATIONS

Your goal is to repeatedly brew the best beer possible. So how do you get there? Use process records to set up product specifications for both in-process and finished beer. This keeps you honest and helps you set an achievable target. Product specifications should consist of both numeric measurements (and their allowable deviations) and a sensory true-to-brand description.

The following specifications (tables 7.1–7.3, fig. 7.1) are based on historic data gathered from one brand of beer and should be used as an example only. Your specifications should be carefully considered based on each of your beers, measurements, process preferences, and capabilities.

TABLE 7.1 SAMPLE BREWHOUSE IN-PROCESS SPECIFICATIONS

Brand X	Parameter measured							
	Mash pH	Mash temp(°F)	Final runnings' SG/°P	Final runnings' pH	Boil SG/°P	Boil pH	Knockout SG/°P	Knockout pH
Target	5.4	154	1.0125 / 3.2	5.65	1.0564 / 13.9	5.30	1.060 / 14.7	5.20
Acceptable deviation +/-	0.12	1.5	0.003 / 0.8	0.14	0.0025 / 0.6	0.15	0.001 / 0.3	0.05

TABLE 7.2 SAMPLE FERMENTATION CELLAR SPECIFICATIONS

Brand X fermentation curve			
Day	Avg. SG/°P	SD (+/-) error bars	pH
0	1.0600/14.7	0.0010	5.19
1	1.0482/12.0	0.0067	4.64
2	1.0278/7.0	0.0069	4.50
3	1.0148/3.8	0.0028	4.51
4	1.0134/3.4	0.0013	4.60
5	1.0130/3.3	0.0011	4.57
6	1.0126/3.2	0.0011	4.58
7	1.0125/3.2	0.0010	4.58
8	1.0127/3.2	0.0011	4.59
9	1.0125/3.2	0.0010	4.61
10	1.0126/3.2	0.0011	4.74
11	1.0119/3.0	0.0012	4.81
12	1.0113/2.9	0.0010	4.82
13	1.0110/2.8	0.0011	4.84
14	1.0108/2.8	0.0011	4.84

HISTORIC BRAND X FERMENTATION CURVE

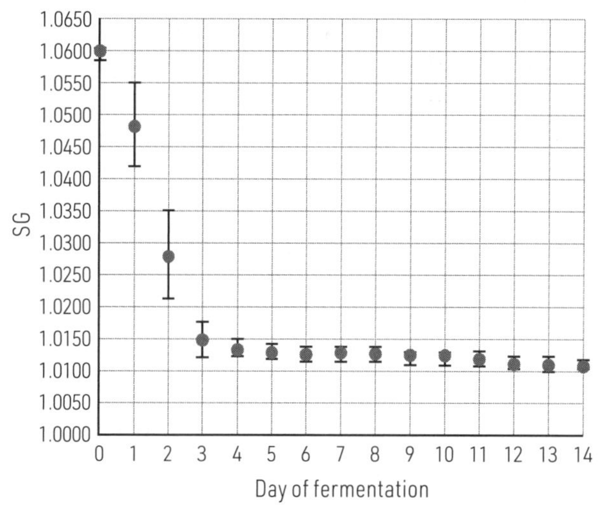

Figure 7.1. Specific gravity against time. Based on data in table 7.2.

TABLE 7.3 SAMPLE BRITE BEER AND PACKAGING SPECIFICATIONS

Brand X	Parameter measured							
	Transfer cells/mL	Package SG	Package pH	Package cells/mL	vol. CO_2	ADF	RDF	ABV %
Target	1.02×10^6	1.0099	4.89	8.50×10^5	2.55	82.59	68.73	6.30
Acceptable deviation +/-	4.32×10^5	0.0011	0.07	3.32×10^5	0.10	1.98	1.59	0.3

ABV, alcohol by volume; ADF, apparent degree of fermentation; RDF, real degree of fermentation

Finally, we add a true-to-brand description. We will discuss developing one of these in chapter 13.

Brand X
IPA 6.3%

Color and appearance
Beautiful color, rich golden amber. Some white head. "Orange tootsie pop."

Aroma
Tons of candied orange peel and grapefruit. Orange, tangerine, cantaloupe.

Some piney resin, evergreen, juniper, bordering on slight thyme herbal notes.

Flavor and aftertaste
Lots of everything carrying over from the aroma!

Big sweet tangerine, orange, grapefruit, pine, a tiny amount of caramel.

Mouthfeel and body
Responses ranged from light to medium. Rounded. Stays fresh and not too sticky. Some noted a very light astringency on the finish.

Carbonation
Tiny bubbles, substantial but not prickly.

Bitterness
Firm and assertive enough to balance body. Lots of grapefruit pith.

Alcohol
Responses ranged from "not perceived" to "slightly warming."

CONSUMER AND IN-HOUSE COMPLAINTS

It is important to keep a log of all consumer complaints. Be sure to note the brand and track how the problem was resolved. If there was a root cause investigation, include the results and findings. In-house complaints are similar, but are more likely to be quality concerns brought up by staff that concern issues such as a kegged beer turned hazy or the last batch smelled a bit eggy. You can use this information log to catch trends with brands or processes.

It is helpful to track the pertinent information on a spreadsheet, that way it will be easier to spot a trend when you notice the same complaint appear multiple times. Trends can be indicators of flaws in your process and can trigger process improvement. Presenting well-documented data such as this can help influence decision-makers to make a capital investment into process improvements.

CORRECTIVE ACTION PLAN

Let's face it, something is going to go wrong eventually. By writing a quality manual you will know who is in charge when something goes wrong. You can put a policy in place that details the corrective actions to take if beer is out of specification or poses a food safety risk. If the product is truly out of control, who makes the final call on destruction? What happens when there is a consumer complaint? Outlining a general protocol that addresses what to do in these cases will help you and your team swiftly decide on the best course of action.

But you are not done there. How do you keep it from happening again? Do a root cause analysis (see chap. 22). Remember, continuous improvement is the name of the game!

ROOT CAUSE ANALYSIS

Root cause analysis is the process of identifying and correcting the underlying cause of a quality issue. If the true root cause is identified and corrected, the quality issue should not manifest itself again. There are many methods and tools available to conduct a root cause analysis. We cover two powerful tools, fishbone diagrams and the Plan-Do-Study-Act Cycle, on pages 196-198 in section 4 of this book.

GOOD LABORATORY PRACTICES

In the good lab practice section of your quality manual you lay out basic rules for how the lab will function on a day-to-day basis. Depending on the space you are working with, there may be limits on how far you can implement some of these practices. That's OK. Most craft brewery labs start in what are essentially closets. You can make it work, just understand your limitations and get creative. Even in a closet, it is best practice lay out some basic rules regarding safety and workflow. If you are having trouble getting started, head over to the ASBC website for their lab safety check list (https://www.asbcnet.org/lab/safety/Pages/default.aspx). It is available to everyone regardless of whether you are a member. But let's be real, you probably should become a member.

Rising Tide's lab is a common workspace; it is used by lab, brewhouse, and production staff for at-line quality measurements as well as product and ingredient analysis. The lab's small footprint makes cleanliness and organization paramount to its daily success. It is the responsibility of each staff member to clean and return all lab tools and glassware to their designated place immediately after finishing their task. This is most important for the DO meter because it is a sensitive and costly piece of equipment. A comprehensive list of good laboratory practices can be found at the back of this quality manual. The following is a brief list of lab polices.

- Because there are harmful chemicals present, eating or drinking is not allowed in the lab.
- Food is not to be stored in the lab refrigerator.
- Prepare samples for consumption in designated lab glassware (ex. "D-test only").
- Lab glassware is not to be used for tasting out of. All blends for tasting are poured into proper serving glassware.
- Mouth pipetting is not allowed.
- Safety data sheets (SDS) for lab chemicals are kept in the lab SDS binder on the bookshelf. All lab-trained employees must read the SDS manual annually.

CALIBRATION PLAN

Brewers rely on many instruments at every step in their process. Scales, gauges, thermometers, pH meters, hydrometers, and many other sensitive pieces of equipment. Most instruments can be calibrated, all instruments can be checked. Lay out a plan for calibration, making a table and listing each instrument on it. For each instrument fill out the following fields:

- Instrument
- Calibration type
- Parameters to check
- Equipment/standards required
- Tolerance
- Frequency

If the instrument cannot be calibrated, it should at least be checked against a known sample value. Refer to the owner's manual for most instrument calibration parameters. Keeping up with instrument calibration is the single most important quality assurance task for any brewer. There is nothing worse than relying on faulty or poorly calibrated instrumentation when trying to make important decisions regarding your beer.

Keeping up with instrument calibration is the single most important quality assurance task for any brewer.

By writing your quality manual you are setting clear expectations for yourself and your staff, and you now have an action plan going forward. It also means you have documented your intentions and the systems used to make your beer. This will go a long way with an FDA auditor if one should show up. As you improve your process, keep adding to the quality manual. Let it be your guide to making pint after pint of quality beer that you can be happy with.

Do not let perfect be the enemy of good. The worst thing you can do is not start your QA/QC program because you cannot make it perfect. Keep in mind that all these components tend to develop simultaneously and not linearly as they are written. Do your best to get it all in writing. Documentation is crucial. If it is not written down (or logged) it likely will not be done. With a solid foundation in place, it is time to start thinking not only about parameters that really matter, but how to measure them. This will put you on our way to making great beer.

PART III
MEASUREMENTS

Measurements make up the mortar that holds the quality priority pyramid together. You use measurements to ensure process controls (CCPs and QCPs) that you set up for your GMP, food safety, and quality plans are working. You also use measurements to set standards for each brand or process. You have already seen an example of this in the "Product Specifications" section of chapter 7. These measurement standards will be used throughout your brewing process to inform you about the condition of your brew and whether or not it is in control.

Part of setting up standards is not just looking at a number but understanding the principles of the measurement and how it affects your process. For each topic in this section we will briefly cover the theory behind the measurement, some possible implications for your process, selected methods, and the tools available and their basic principles. All of this is to help you really understand what the numbers mean for you so that you can choose the standards that are correct for your brewery. At the end of each chapter there is a brief list of QCPs you may decide to add to your quality plan and start measuring.

The measurement chapters are organized in general order of priority for sub-10,000-barrel craft breweries. **Priority 1** measurements are for breweries who for sell beer for consumption on the premises only. That means even the smallest brewery operation should be making these measurements for every batch brewed. **Priority 2** measurements should be made (in addition to the priority 1 measurements) by breweries who are selling beer that may be consumed outside of their own brewery walls. **Priority 3** measurements are made when you are looking to fine-tune your processes, but only after priority 1 and 2 measurements are well monitored and controlled. Of course, these are generalities and you may shift priorities up or down depending on what is right for your operation.

PRIORITY 1

- Weight and volume
- Density
- Temperature
- pH
- Yeast
- Sensory analysis

PRIORITY 2

- Microbiology
- Package quality measurements
- Carbon dioxide
- Dissolved oxygen
- Alcohol
- Titrations

PRIORITY 3

- Refine priority 1 and 2 measurements
- Spectrophotometry

8

USEFUL CALCULATIONS AND EQUATIONS

The brewer looking to achieve quality and consistency needs to be able to accurately measure liquid volumes. Of equal importance is being able to determine tank capacities when blending, adding fruit, or testing a recipe with an unusual or extreme malt bill. And in the lab, understanding weight and volume is important when preparing solutions of known concentration; there are many cases when these concentrations need to be scaled up to the size of the brewhouse. In the brewery lab there is no escaping algebra. I will introduce some of the most commonly used equations and problem-solving techniques used in the brewery lab. I know it has been a while since I have taken a math class, so let us start with something familiar and work our way through them. You can do this!

CALCULATIONS INVOLVING VOLUME

Basic Volume Calculations

Volume of a rectangular vessel. Perhaps the simplest volume calculation is that for a rectangular vessel. This takes into account the three dimensions of length, width, and height. For example, let us look at the volume of a rectangular vessel such as a coolship that is 8 ft. by 5 ft. with a height of 3 ft.

> **UNITS**
>
> It goes without saying, but I will say it anyway: throughout this book, references to gallons (gal.) mean US gallons (1 gal. = 3.785 L) and barrels (bbl.) mean US beer barrels (1 bbl. = 31 gal.). Other non-metric units, such as feet (ft.) and inches (in.), also follow the US customary system, although it is usually only units of volume where differences arise between different non-metric systems.
>
> Metric units are the same the world over.

$$\text{length} \times \text{width} \times \text{height} = \text{volume (cubic units)} \tag{8.1}$$

$$8 \text{ ft.} \times 5 \text{ ft.} \times 3 \text{ ft.} = 120 \text{ ft.}^3 \tag{8.2}$$

OK, we know how many cubic feet the coolship occupies, but how can we express how much beer will it hold in units that are more useful? From the chart in appendix F we can see 1 US beer barrel (bbl.) occupies 4.144 ft.^{3}. Here's that math in all its glory:

$$\frac{1 \text{ bbl.}}{4.144 \text{ ft.}^3} = 1$$

In other words, if we use the ratio of barrels to cubic feet as a conversion factor, then

$$\frac{4.144 \text{ ft.}^3}{1} \times \frac{1 \text{ bbl.}}{4.144 \text{ ft.}^3} = 1 \text{ bbl.} \tag{8.3}$$

From equation (8.2) we see the coolship is 120 ft.^{3}, which can be converted to barrels using the conversion factor in (8.3):

$$\frac{120 \text{ ft.}^3}{1} \times \frac{1 \text{ bbl.}}{4.144 \text{ ft.}^3} = 28.96 \text{ bbl.} \tag{8.4}$$

This means 28.96 bbl. will fill the coolship to the very top, if the coolship is on a level surface and the temperature of the wort is 68°F that is. We will discuss the liquid volume and temperature relationship in the next section.

Volume of a cylinder. Being able to calculate the volume of a cylinder allows you to calculate the volume in a cylindrical tank, pipe, or hose.

$$\pi \times r^2 \times \text{height (or length)} = \text{volume (cubic units)} \tag{8.5}$$

Let us say we are setting up to clean a fermentor with 20 ft. of transfer hose with a 1.5 in. internal diameter. To ensure the correct dose of chemical is added, we should include the amount of water in the hose. First, convert the diameter in inches to feet, 1.5 in. × (1 ft./12 in.) = 0.125 ft. To be able to use (8.5), the radius, r, can be derived from the diameter because $r = \frac{1}{2} d$, hence, the radius of the hose is calculated as ½ (0.125 ft) = 0.0625 ft. Plugging that all into equation (8.5) gives

$$3.14 \times (0.0625 \text{ ft.})^2 \times 20 \text{ ft.} = 0.2454 \text{ ft.}^3 \tag{8.6}$$

One more step: from equation (8.3) we know 1 bbl. occupies 4.144 ft.^{3}, so if we want to express the volume in the hose in gallons we convert

$$\frac{4.144 \text{ ft.}^3}{1 \text{ bbl.}} \times \frac{1 \text{ bbl.}}{31 \text{ gal.}} = \frac{0.1337 \text{ ft.}^3}{\text{gal.}} \tag{8.7}$$

For convenience, table 8.1 includes volumes/capacities for hose of varying internal diameter as a quick "cheat sheet."

TABLE 8.1 VOLUME OF LIQUID PER FOOT OF HOSE ACCORDING TO INTERNAL DIAMETER

Internal diameter (in.)	Gallons of liquid per foot
1.0	0.0408
1.5	0.0918
2.0	0.1632
2.5	0.2550
3.0	0.3672
4.0	0.6528

From (8.6), we know the hose holds 0.2454 ft.^{3}, thus

$$\frac{0.2454 \text{ ft.}^3}{0.1337 \text{ ft.}^3/\text{gal.}} = 1.835 \text{ gal.} \tag{8.8}$$

Volume of a cone. For the sake of completeness, let us calculate the volume of a cone. Knowing all these volumetric calculations should enable you to determine the volumes and capacities of just about any vessel you come across. The equation for the volume of a cone is

$$\frac{\pi \times r^2 \times \text{height}}{3} = \text{volume (cubic units)}. \tag{8.9}$$

Example

Let us work through an example that puts all of the above together. In this example, suppose you overshot your density on brew day and must add water to the brew kettle to dilute back to your target OG. The question you need to answer is, will it all fit in the fermentor after knockout? Here is what you know:

- total tank height is 6 ft.
- cone height is 1.5 ft.
- radius is 2 ft.
- knockout is 16 bbl. of wort
- you need to add 3 bbl. of water to hit the target OG

Your steps are to

1. use equation (8.9) to determine the volume of the cone;
2. use equation (8.5) to determine the volume of the cylindrical tank body; and
3. total the volumes and divide by the volume of 1 bbl. (eq. 8.3).

Step 1, determine the volume of the cone:

$$\text{volume of cone} = \frac{\pi \times r^2 \times \text{height}}{3}$$

$$= [3.14 \times (2 \text{ ft.})^2 \times 1.5 \text{ ft.}] / 3$$
$$= 18.84 \text{ ft.}^3 \quad (8.10)$$

Step 2, determine the volume of the cylinder:

$$\text{volume of cylinder} = \pi \times r^2 \times \text{height cylindrical tank body}$$
$$= 3.14 \times (2 \text{ ft.})^2 \times (6 \text{ ft.} - 1.5 \text{ ft.})$$
$$= 56.52 \text{ ft.}^3 \quad (8.11)$$

For the first part of step 3, calculate the total volume of the tank by adding the two volumes from (8.10) and (8.11) together:

$$18.84 \text{ ft.}^3 + 56.52 \text{ ft.}^3 = 75.36 \text{ ft.}^3 \quad (8.12)$$

You can now convert the total capacity in (8.12) from cubic feet to US beer barrels:

$$\frac{75.36 \text{ ft.}^3}{1} \times \frac{1 \text{ bbl.}}{4.144 \text{ ft.}^3} = 18.19 \text{ bbl.} \quad (8.13)$$

The 18.19-barrel capacity is less than the total of the knockout wort and water needed to add to hit your target OG. At this point you have a decision to make. You can either knockout the full volume of high-gravity wort and accept that you missed the target, or you can adjust the knockout and water volume to hit the target and not overfill the fermentor. There are many surrounding factors that will influence which decision you make here. I am going to assume that this is a brand with target specifications so I will elect to choose the second option.

In this example, you know that adding 3 bbl. of water to 16 bbl. of knockout wort will hit the target OG for this brand. Since your tank only holds 18.19 bbl, we will choose to knockout 17 bbl. to leave a little headspace for fermentation. Use the ratio of added water to total volume (eq. [8.14]) to determine how much water to add to make 17 bbl. total volume (eq. [8.15]). From this, you will determine how much wort to knockout (eq. [8.16]).

$$\frac{3 \text{ bbl.}}{3 \text{ bbl.} + 16 \text{ bbl.}} = 0.157 \quad (8.14)$$

$$0.157 \times 17 \text{ bbl.} = 2.67 \text{ bbl. of water to add} \quad (8.15)$$

$$17 \text{ bbl.} - 2.67 \text{ bbl.} = 14.33 \text{ bbl. of KO wort} \quad (8.16)$$

So, now you have determined that you can knockout 14.33 bbl. of wort and add 2.67 bbl. of water to hit your target OG for the brand and avoid overfilling your fermentation vessel.

Volume and Temperature

Now that we can calculate volumes of many shapes of vessels, we must consider the physical behavior of liquids when heated. As energy is added to a liquid (i.e., heating) the molecules of the liquid start vibrating faster and the temperature rises, in other words, the average kinetic energy of the molecules in the liquid increases. As the molecules vibrate more the total volume occupied by the liquid increases. This expansion in volume is directly related to the increase in temperature.

As mentioned after equation (8.4), the baseline temperature for measuring volumes of liquid in brewing is 68°F. The standard rule in brewing is that wort cooled to 68°F (20°C) occupies only 96% of the volume of the same mass of boiling wort at 212°F (100°C).[1] Note that the mass of the wort stays the same because the amount of wort does not change, just the volume it occupies. If you are chasing a discrepancy in IBU or yield, try looking at the 4% contraction of volume between boiling and cool liquid.

$$\text{volume of boiling wort (212°F)} \times 0.96$$
$$= \text{volume of cool wort (68°F)} \quad (8.17)$$

Volume, Density, and Weight

There are many instances in the brewery where volume can be measured by weight. To do this we take advantage of the density measurement of liquids. Density is mass per unit volume, and if you are keen on your dimensional analysis (see next section) you can see where this is going rather quickly. I will go more in-depth on density in later chapters; for now, you need to know that water at 68°F weighs approximately one gram per cubic centimeter (1 g/cm^3). This fact can be scaled to larger volumes (see table 8.2).

[1] Steven R. Holle, *A Handbook of Basic Brewing Calculations* (St. Paul, MN: Master Brewers Association of the Americas, 2003, 2010), 14.

TABLE 8.2 WEIGHT OF WATER AT 68°F BY COMMON UNITS OF VOLUME

Volume	Mass
1 mL	1 g
1 L	1 kg (2.2 lb.)
1 gal.	8.32 lb.
1 bbl.	258 lb.

When we dissolve malt extract in water, the density of the water will increase, hence, the mass of the extract solution increases relative to water. Wort or beer (or any other solution) with a specific gravity (SG) of 1.015 (3.8°P) is actually 1.5% heavier than water, which has an SG of 1.000 (0.0°P). So, if you know the density of a liquid, you can derive how much the desired volume should weigh. This eliminates the need to visually see the fill volume.

$$\text{weight of solution} = \frac{\text{mass}}{\text{volume}} \times \text{SG} \times \text{volume of beer} \quad (8.18)$$

As an example, you can use equation (8.18) to determine the weight of 5.15 gal. of 1.011 SG beer in a keg.

$$\begin{aligned}\text{weight of solution} &= \frac{\text{mass}}{\text{volume}} \times \text{SG} \times \text{volume of beer} \\ &= (8.32\text{ lb.}/1\text{ gal.}) \times 1.011\text{ SG} \times 5.15\text{ gal.} \\ &= 43.32\text{ lb.} \quad (8.19)\end{aligned}$$

DIMENSIONAL ANALYSIS AND UNIT CONVERSIONS

Dimensional analysis. Sounds fancy, right? Do not worry, I have already snuck one or two examples of dimensional analysis into the calculations we have looked at, I just have not defined it yet. Dimensional analysis is a general problem-solving approach that uses the dimensions or units of each value to guide you through the calculation. To do this, think of the units as algebraic entities (e.g., ft. × ft. = ft.^{2}) and ensure you follow the units attached to your measurements through your calculations. In the following, the algebraic entities x and y represent units:

$2x + x = 3x$	2 ft. + 1 ft. = 3 ft.
$2y - y = y$	2 in. – 1 in. = 1 in.
$2x \cdot x = 2x^2$	2 ft. × 1 ft. = 2 ft.^{2}
$\frac{2y^2}{y} = 2y$	$\frac{2\text{ in.}^2}{1\text{ in.}} = 2\text{ in.}$
$\frac{3x}{36y} = \frac{x}{12y}$	$\frac{3\text{ ft.}}{36\text{ in.}} = \frac{1\text{ ft.}}{12\text{ in.}}$

A key step in dimensional analysis is making sure you are working in the same units. The easiest way to do so is to use a conversion factor. A conversion factor is a ratio between one unit of measure and another. Because the numerator and the denominator in the conversion describe the same quantity only in different units, the conversion factor is an algebraic entity that equals 1. In general, a conversion factor is used thus:

$$x\text{ in original unit}\left(\frac{\text{new unit}}{\text{original unit}}\right) = x\text{ in new unit} \quad (8.20)$$

That is exactly what online converters do when you convert any unit to another. I used this method twice when working out the volume of water in a cylinder earlier in this chapter. In that example above, the cylinder was a hose with the diameter in inches, which we converted into feet using a conversion factor:

$$1.5\text{ in.}\left(\frac{1\text{ ft.}}{12\text{ in.}}\right) = 0.125\text{ ft.} \quad (8.21)$$

Let us look another example, this time moving from US customary units to metric using the same method. In this case, we will take the answer from (8.21) and convert it to centimeters

$$0.125\text{ ft.}\left(\frac{30.48\text{ cm}}{1\text{ ft.}}\right) = 3.81\text{ cm} \quad (8.22)$$

For the sake of completeness let us work through converting the volume of a US beer barrel to metric. Suppose you know there are 3.785 liters (L) in a gallon, but you cannot remember how many liters are in a cubic foot. However, you know from (8.3) how many cubic feet are in a barrel and you know from (8.7) how many gallons are in a cubic foot. By chaining the conversion factors for each, you can arrive at the number of liters in a barrel. Remember that by treating units as algebraic entities (x, y, etc.) they cancel in the resulting expression, leaving you with the units you are trying to convert to.

1 bbl. in liters = 1 bbl. × (ratio ft.^{3} to 1 bbl.) × (ratio gal. to ft.^{3}) × (ratio L to gal.)

$$= 1\text{ bbl.}\left(\frac{4.144\text{ ft.}^3}{1\text{ bbl.}}\right)\left(\frac{1\text{ gal.}}{0.1337\text{ ft.}^3}\right)\left(\frac{3.785\text{ L}}{1\text{ gal.}}\right)$$

$$= \frac{1 \times 4.144 \times 1 \times 3.785\text{ bbl.ft.}^3\text{gal.L}}{1 \times 0.1337 \times 1\text{ bbl.ft.}^3\text{gal.}}$$

$$= \frac{15.685\text{ L}}{0.1337}$$

$$= 117.3\text{ L} \qquad (8.23)$$

Dimensional analysis is a powerful method and comes up repeatedly. It has many uses in both the lab and the brewery.

CHEMISTRY TERMINOLOGY REFRESHER

Standard atomic weight: can be found under the element symbol on the periodic table, sometimes labeled $A_{r,\,std}$. This number represents the weighted average atomic mass of an element. It is an average because elements naturally occur as different isotopes that have slightly different relative atomic masses.

Isotopes: atoms of the same element that differ in mass due to differences in neutron number. All atoms of an element have the same number of protons in the nucleus (its *atomic number*) but they may differ in the number of neutrons.

Mole (mol): a unit of measurement for the amount of a substance. One mole is equal to $6.02214076 \times 10^{23}$ particles of a substance, whether that be atoms, molecules, or ions. One mole of substance A contains the same number of particles as 1 mol of substance B, although their masses may be very different.

Molar mass: the mass in grams of 1 mol of a substance.

Molarity (M): moles of solute per liter of solution (mol/L). For example, if 1 mol of substance A is dissolved in 1 L of water then the concentration is 1 mol/L or 1 M. You would say it is a "one-molar solution."

WORKING WITH SOLUTIONS

All around the brewery we are preparing solutions of know concentrations, whether it is extract in the brewhouse, desired concentrations of cleaning chemicals, or a precise molar quantity of chemical for quantitative measurements in the lab. In general, there are two ways this is done: combining a weighted solute with a solvent, or diluting a more concentrated solution. To get you through these two methods I will use an example of making a stock solution and then an example of diluting that stock solution down to a working concentration. Put on your lab coats because we are going to do some chemistry here.

Making Standard Solutions

Example 1: weighted solute with solvent

Suppose you have reagent grade sodium hydroxide (NaOH) pellets and want to make 2 L of a 1-molar NaOH stock solution in preparation for a series of total acidity titrations.

Determine the moles of NaOH required to make 2.0 L of 1 M NaOH.

$$\frac{2.0\text{ L} \times 1\text{ mol NaOH}}{1\text{ L}} = 2\text{ mol NaOH required} \qquad (8.24)$$

Determine the mass of 2 mol NaOH. You can consult the periodic table or list of elements to find the standard atomic weight of each element, in this case, sodium (Na), oxygen (O), and hydrogen (H). A molecule of NaOH has a total molecular weight of 39.9971, which means 1 mol NaOH weighs 39.9971 g.

$$\frac{2\text{ mol NaOH} \times 39.9971\text{ g NaOH}}{1\text{ mol NaOH}} = 79.9942\text{ g NaOH} \qquad (8.25)$$

Considering the precision of your lab equipment, dissolve 79.99 g of NaOH in enough water to make 2 L of solution. *Note*: you do not add 79.99 g to 2 L. Instead, dissolve the NaOH in a smaller volume of distilled water in a volumetric flask and top up with distilled water to 2 L. If you do not have a volumetric flask you can accomplish the same by adding the 79.99 g NaOH to a total of 1,920.1 g of distilled water for a final volume of 2 L (this takes advantage of the fact that 1 mL of water weighs approximately 1 g at 68°F).

The result is a 1 M NaOH stock solution. Method one complete!

When mixing acids or bases with water be sure to add the chemical to a volume of water, NOT water to concentrated chemical. A lot of reactions with water are exothermic (give off heat). If the reaction is not controlled by slowly adding the chemical to water it can lead to boiling and bubbling, which can cause dangerous chemicals to spit out of the container.

Example 2: diluting a concentrated solution
Suppose on the day you will be titrating you estimate you will need 500 mL of 0.1 M NaOH to complete all the titrations. You want to dilute your NaOH stock solution from 1.0 M to 0.1 M. What volume of 1.0 M NaOH solution must be diluted to make 500 mL of 0.1 M NaOH solution?

Determine the amount of NaOH in dilute solution. From above, we know the desired volume (V) is 500 mL and the desired concentration (C) is 0.1 M.

$$\text{amount(moles) of NaOH needed dilute solution} = C_{\text{desired}} \times V_{\text{desired}}$$

$$= \frac{0.1\ \text{mol}}{1\ \text{L}} \times 0.5\ \text{L}$$

$$= 0.05\ \text{mol} \qquad (8.26)$$

Determine the amount of stock solution needed to contribute 0.05 mol NaOH.

$$\text{vol. 1 M NaOH needed} = 0.05\ \text{mol NaOH} \times \frac{1}{1\ \text{mol NaOH}}$$

$$= 0.05\ \text{L} = 50\ \text{mL} \qquad (8.27)$$

Add 50 mL of your 1.0 M stock solution to a volumetric flask or beaker and top up with distilled water to a total volume of 500 mL. This will give you 500 mL of 0.1 M NaOH titrant solution.

Serial Dilutions

Serial dilutions are used to accurately and reliably reduce the concentration of a substance in solution to some lower concentration. This method is often used to reduce a highly concentrated stock solution of chemical to build a concentration curve or to dilute a yeast slurry for analysis under the microscope. No matter what you are trying to dilute the procedure is the same.

- Take 1 mL of concentrated solution and add 9 mL of distilled water for a total of 10 mL. This is a 1:10 or 10× dilution.
  - For example, 1 mL of 1 M NaOH diluted with 9 mL distilled water is a 10× dilution. The resulting concentration is 0.1 M NaOH.

Depending on your needs you may choose to dilute this concentration further.

- Take 1 mL of the 10× diluted solution and add 9 mL of distilled water for a total of 10 mL. This is a 1:100 or 100× dilution.
  - For example, further dilution of 1 M NaOH by taking 1 mL of the 0.1 M diluted solution and adding 9 mL distilled water is a 100× dilution. The resulting concentration is 0.01 M NaOH.

Using this series of steps, you will be diluting by a factor of 10 each time. The original 1 M solution is diluted to 0.1 M, 0.01 M, 0.001 M, and so on. If using this method to dilute and count yeast you can calculate back to the original concentration of the slurry by multiplying by the dilution factor.

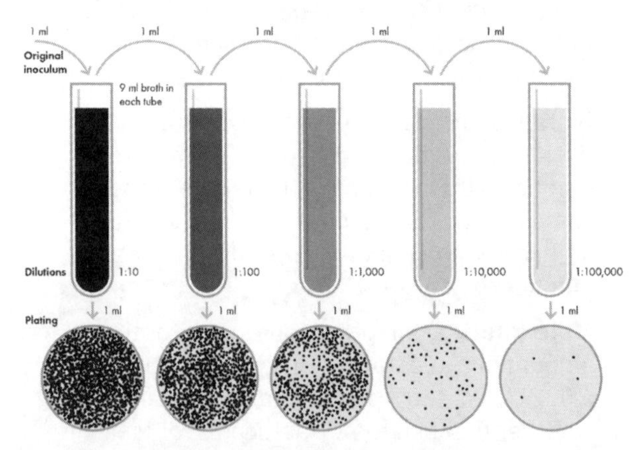

Figure 8.1: The serial dilution of a dense yeast sample and the resulting colony forming units (CFUs) cultured on growth media.

Mixing Equation

The mixing equation is immensely helpful around all parts of the brewery. It allows you to calculate a blend from known concentrations (C_1, C_2, . . .) and volumes (V_1, V_2, . . .). You can also do some algebra to hit a desired target (C_t) from known concentrations.

$$C_1V_1 + C_2V_2 = C_t(V_1 + V_2) \qquad (8.28)$$

This equation can be useful for many things, including pitching yeast slurries or blending for alcohol content. It can be used not only for concentrations and volumes but any quantity and its associated characteristic.

As an example, suppose you blend two oak barrels of beer of known ABV and you want to know the resulting ABV.

- Oak barrel one: 53 gal. at 8% ABV
- Oak barrel two: 49 gal. at 11% ABV

Rearrange equation (8.28) to isolate C_t:

$$C_t = \frac{C_1 V_1 + C_2 V_2}{V_1 + V_2} \tag{8.29}$$

Solve for C_t by plugging the values into (8.29):

$$C_t = \frac{(0.08 \times 53) + (0.11 \times 49)}{53 + 49}$$

$$= (4.24 + 5.39) / 102$$
$$= 0.094, \text{ or } 9.4\% \text{ ABV} \tag{8.30}$$

Another example of using the mixing equation would be to dilute a concentrated stock solution. Suppose you have a 1% stock solution of methylene blue and want to make 500 mL of 0.01% methylene blue. You can use a simplified version of the mixing equation

$$C_1 V_1 = C_t V_t$$

where

C_1 = 1%
C_t = 0.01%
V_t = 500 mL

Insert the values into equation (8.30):

$$1\% \times V_1 = 0.01\% \times 500 \text{ mL}$$

Solve for V_1:

$$V_1 = \frac{0.01\% \times 500 \text{ mL}}{1\%}$$

$$= 5 \text{ mL}$$

This shows that you need 5 mL of stock 1% methylene blue to be added to 495 mL of distilled water for a total of 500 mL of 0.01% working strength methylene blue.

Scaling

The scaling equation is one final, helpful equation that allows you to scale up or down concentrations if you have a recipe for a large volume and would like to scale that back to a smaller volume. For instance, suppose the instructions on your HLP media call for 7 g of powdered media per 500 mL of distilled water. In order not to waste media, you want to scale that down to 200 mL, for which you need to determine how many grams of media to add.

$$\frac{\text{original quantity}}{\text{original volume}} = \frac{\text{unknown quantity } x}{\text{known desired volume}} \tag{8.31}$$

It is important to note that the units used for the ratios should match.

$$\frac{\text{original quantity (grams)}}{\text{original volume (mL)}} = \frac{\text{unknown quantity } x \text{ (grams)}}{\text{desired volume (mL)}}$$

Now the equation can be rearranged to solve for "unknown quantity x (grams)," which we can simply call x.

$$x = \left(\frac{\text{original quantity}}{\text{original volume}}\right) \cdot (\text{desired volume}) \tag{8.32}$$

Now plug the values into (8.32). Here, x will equal the new quantity of media in grams:

$$x = \left(\frac{7 \text{ g}}{500 \text{ mL}}\right) 200 \text{ mL}$$
$$= 2.8 \text{ g}$$

This equation can be applied to many linear scaling relationships.

While I promise this is not a math textbook there will be more equations presented as needed, but this was by far the most equation-dense chapter. The good news is, the equations presented here are some of the most common equations used in the brewery and you have them at your fingertips. They will work on a large scale for wort production and blending as well as the small scale in the lab. They are the basis of advanced analysis techniques such as yeast counting, reagent preparation, and much more.

9

DENSITY

I remember opening my first homebrew kit. Everything I needed to brew was right there at my fingertips. Once the equipment was all set up, I felt one-part alchemist, one-part mad scientist, especially when I fixated on this weird looking cylinder with a smaller cylinder on top. What could this be for? I came to learn it was a hydrometer and just about every homebrewing starter kit comes with one. And for good reason: measuring the density of wort and beer is quite possibly the most informative measurement a brewer can make over the life cycle of their beer. Measuring the **original extract** (OE), often referred to as original gravity (OG), allows you to calculate extract yield of your mash and from there you can calculate brewhouse efficiency. By tracking gravity, you know how much extract is being consumed by the yeast throughout fermentation; you can use the difference between density at the beginning of fermentation and the end of fermentation to get a general idea of the ABV of your beer.

DEFINING DENSITY

Density (ρ) is a property of matter and is defined mathematically as mass per unit volume,

$$\rho = \frac{m}{v},$$

where ρ is the density, m is the mass, and v is the volume. This gives density the SI-derived unit of kilograms per cubic meter (kg/m^3), but it is common to see grams per cubic centimeter (g/cm^3) used in many technical fields. You may be saying to yourself, "My hydrometer reads a different scale, specific gravity or degrees Plato, not grams per cubic centimeter." Before we discuss these other scales that are used daily, it is important to understand density, its limitations, and how it relates to what we are measuring.

To begin, it is helpful to think about density on the atomic scale. The mass of atoms, their size, and how they are arranged determine the density of a substance. Let us conduct a mini-thought experiment. Imagine you are holding a 1 cm^3 block of stainless steel in your right hand and a 1cm^3 block of wood in your left hand. I am sure you would agree that the volume is exactly the same but that the mass of the stainless-steel cube is greater as it feels heavier in your hand. Using the equation above, let us roughly compare the two substances:

$$\rho_{\text{stainless}} = \frac{\text{a lot (g)}}{1\ \text{cm}^3} \qquad \rho_{\text{wood}} = \frac{\text{a little (g)}}{1\ \text{cm}^3},$$

therefore,

$$\rho_{\text{stainless}} > \rho_{\text{wood}}.$$

It is clear that the density of stainless steel is greater than the density of wood as long as we hold the volume fixed. The mass of the stainless-steel cube is greater because the iron, carbon, chromium, and nickel atoms are more tightly compacted than that of the carbon, hydrogen, and oxygen atoms of the wood cube.

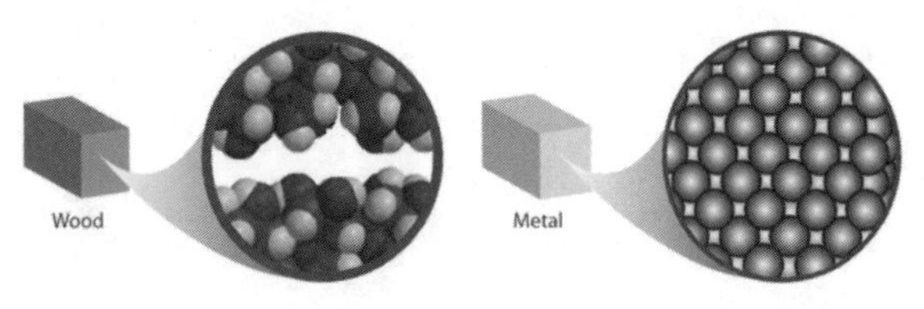

Figure 9.1. An illustration of differences in the compaction of atoms for wood and metal.

In general, we are not concerned with the density of solid substances in the brewhouse. But, just like solids, the mass and size of the atoms in a liquid and how closely they are packed together determine the liquid's characteristic density. The density of liquid is determined using the exact same equation above. Let us take this concept into the brewhouse. During the mashing process we are extracting sugars, dextrins, proteins, and other substances, increasing the density of our brewing water and creating the solution brewers call wort. Looking at this process in terms of density, we are starting with water, which is known to have a density of approximately 1.00 g/cm^3. We then dissolve solids (our extract) into the liquid, increasing the mass of the liquid. Thus, the resulting liquid solution, the wort, has a density greater than water, probably around 1.05 g/cm^3. Since the density of a substance is the same regardless of its sample size, you can think of this as adding 0.05 g of molecules per 1 cm^3 of brewing water to create your wort.

There are two major factors that affect density: pressure and temperature. Both change the density by changing the volume of the substance. When pressure is applied, the liquid is compressed into a smaller volume and, since the liquid's mass stays the same, its density increases. Luckily for us in the brewery, most density measurements are typically taken in the open air under normal and consistent atmospheric conditions, so we do not have to worry about pressure. The second condition that influences density is temperature. Heating a solution adds energy, causing the molecules to vibrate back and forth to a greater degree, which takes up more room and increases the volume. In this case, the liquid's mass stays the same and its volume increases, reducing the density of the liquid. With this in mind, for consistent results a brewer should take their density readings at the same temperature every time, typically at room temperature, 68°F (20°C).

SPECIFIC GRAVITY AND PLATO

Now that we understand some key factors that influence density, we can look at the scales brewers use to measure density. **Specific gravity** (SG) is defined as the ratio of the density of a substance to the density of a standard, where the standard is water and the substance is beer or wort at the same temperature.

$$SG = \frac{\rho_{\text{substance}}\ (\text{g/cm}^3)}{\rho_{\text{water}}\ (\text{g/cm}^3)}.$$

This relative density does not have units, therefore, the SG of water at 4°C is reported as 1.000, not 1.000 g/cm^3.

Specific gravity readings above 1.000 mean a density greater than water. As discussed earlier, sugars extracted from the mash increase the density of wort, so the SG increases. As yeast consumes sugars in the wort, the SG decreases. The relationship between sugar content and SG is reasonably linear, making SG a useful scale for the brewer to measure extract content. It is worth noting that the mix of compounds dissolved in wort does not change the SG exactly as pure sugar does, but this is ignored by convention. That is, a wort with specific gravity x is assumed to have the same extract content as a pure sugar solution with the same specific gravity.[1]

Possibly the most widely used scale for measuring density by brewers is **degrees Plato** (°P). Degrees Plato expresses the weight percentage of sugar solutions and relates this weight to specific gravity. A solution of 1°P has the same specific gravity as a sucrose solution consisting of 1 g of sucrose dissolved in enough water to make 100 g of solution (1 g sugar to 99 g water). In other words, 1°P is 1% sucrose by weight.[2] Again, temperature is important here because the Plato scale is based on the density of sugar solutions at 68°F (20°C).

After fermentation has started the density measurement in degrees Plato or specific gravity is skewed

1 Kaiser [pseud.], "On the relationship between Plato and specific gravity," *Brewer's Friend* (blog), October 31, 2012, https://www.brewersfriend.com/2012/10/31/on-the-relationship-between-plato-and-specific-gravity/.

2 Karl Ockert, *Fermentation, Cellaring, and Packaging Operations* (St. Paul, MN: Master Brewers Association of the Americas, 2006), 209.

by the presence of alcohol. **Apparent extract** (AE) is the direct measurement of total extract in wort or beer not corrected for alcohol content. Alcohol has a specific gravity substantially lower than water. This means the "true" or **real extract** (RE) is greater than that measured directly.

To convert between °P and SG for apparent extract values, the American Society of Brewing Chemists (ASBC) has published the following equation, which is also used in their online calculator:

$$SG = \frac{°P}{258.6 - [(°P/258.2) \times 227.1]} + 1$$

If you find yourself in a conversation with a brewer who works in Plato while you work in SG, a good approximation between Plato and SG is to treat one degree Plato as equal to four "points" of SG:

$$\text{degrees Plato} = \frac{\text{SG points}}{4}$$

For example, an SG reading of 1.040 is 40 "points"; using the equation above, 40 SG points is approximately 10°P.

HOW TO MEASURE DENSITY

Hydrometer (Saccharometer)

The hydrometer is typically the (home)brewer's first piece of lab equipment used for quality control. It is used to determine the SG (or °P) of liquids. Made of glass, the bulb is filled with a heavy material that keeps the instrument upright when floating in liquid, while the stem of the hydrometer is marked with a scale. The way these instruments work is based on the concept of buoyancy. A hydrometer will float higher in a dense solution and sink further into a less dense solution.

Hydrometer readings are temperature dependent. All hydrometers are calibrated to a certain temperature, typically 59°F (15°C) or 68°F (20°C). A correction scale to account for temperature differences is usually included with a hydrometer and there are various calculators available online to help with this correction. The Brewers Association and the ASBC have collaborated to produce an educational video series for training on the *ASBC Methods of Analysis*, one of which covers "ASBC Beer Method 3b: Apparent Extract Determination using Hydrometers."

It is important to note that after fermentation has started the solution we are measuring is no longer just wort sugars dissolved in water, it now contains ethanol and CO_2. Ethanol is less dense than water, which drives the density down and causes inaccuracies when estimating the amount of extract consumed based on falling density. The result is that the presence of ethanol will make it seem like more sugar has been consumed than is actually the case, hence, *apparent* extract. An estimate of the amount of sugar consumed that is based on apparent extract is called the **apparent degree of fermentation** (ADF) or apparent attenuation. If CO_2 is present in the solution, bubbles will attach themselves to the side of the hydrometer and drive it upward, making the solution seem denser. Degassing the beer is crucial for getting an accurate reading. Pouring the beer back and forth between two vessels is a popular way of degassing and can produce predictable results. Ensure that everyone who is taking a reading degasses using the same technique. The ASBC has published a beer degassing matrix to help with this subject.

Refractometer

Refractometers are popular with brewers because only a few drops of sample are needed to get a reading. A refractometer is a tube resembling a small telescope, with an eyepiece on one end and a prism with a hinged sample chamber at the other. A refractometer measures the density of a solution based on its refractive index, that is, the degree to which a beam of light passing through the test solution is bent from its original path. A refractometer is accurate with water and sugar solutions but has the same limitations as a hydrometer when reading a solution containing ethanol. There are calculators available online that correct a refractometer reading using the OE as a reference point and applying a correction factor to the AE reading of the fermenting wort. Another disadvantage is that solids in solution will also scatter light, making refractometer readings blurry and hard for the user to interpret. Even given these potential inaccuracies, when the refractometer is used in conjunction with a consistent sampling plan useful data can be collected.

TABLE 9.1 OVERVIEW OF DENSITY RELATED TERMINOLOGY

Parameter	Description	Equation
Original extract (OE) (a.k.a. original gravity, OG)	Solids extracted from grist as % wt./wt. It is convenient to use degrees Plato for these equations	$\frac{\text{weight extract}}{\text{total weight wort}} \times 100 = °P$
Apparent extract (AE)	After fermentation has started, AE is the direct measurement of total extract in wort or beer not corrected for alcohol content. Alcohol has a specific gravity substantially lower than water. This means the "true" or real extract (RE) is greater than that measured directly	Direct measure of fermenting wort or beer
Apparent degree of fermentation (ADF) or apparent attenuation	Observed reduction of wort extract not accounting for the density of alcohol in solution	$ADF = 100 \times \left(\frac{OE - AE}{OE}\right)$ AE reading taken at end of fermentation
Real extract (RE)	Total extract in wort corrected for the actual amount of alcohol in the wort. RE calculations account for the presence of alcohol in the finished beer and the absence of alcohol in the starting wort	$RE = (0.1886 \times OE) + (0.8114 \times AE)$ Use °P for OE and AE
Real degree of fermentation (RDF) or real attenuation	RDF is the measured percentage of wort extract that is fermented	$RDF = \left(100 \times \frac{OE - RE}{OE}\right)\left(\frac{1}{1 - (0.005161 \times RE)}\right)$

Digital Refractometer

Like its analog counterpart, a digital refractometer takes measurements based on a sample's refractive index. The difference is that there is an internal light source, and most will have built in automatic temperature correction (ATC). Digital refractometers are typically very affordable, use a small sample size, and remove human error from the interpretation of the result. Similar limitations to those with analog refractometers arise, with the presence of ethanol and light-scattering solids in the wort skewing the results. You will still need to use a calculator or correction factor for readings once fermentation has started.

Digital Density Meter (Densitometer)

Densitometers come in benchtop and handheld models but most work in the same manner. A sample is injected into the machine, where it sits in a U-shaped tube that oscillates. The specific gravity is calculated by comparing the frequency of the oscillating sample liquid to a known standard. Densitometers are the workhorses of the brewery lab. Like hydrometers and refractometers, densitometers measure apparent extract. Densitometers are easy to use, require a relatively small sample size, automatically calculate temperature corrections, and are incredibly accurate between different users. The downside is that they can be clogged easily. This is easy to avoid by running samples through a sieve (150 μm or 150 micron) before injecting them into the densitometer. See your manufacturer's recommendations for recommended sample preparation.

Alcolyzer

An alcolyzer typically consists of a high-quality densitometer and a near-infrared (NIR) spectrophotometer. The NIR spectrophotometer is used to measure alcohol in the sample. Using an NIR measuring method eliminates the influence of other sample constituents on the alcohol measurement, which means a high level of precision can be achieved.[3] While an alcolyzer can be pricey, the device provides a complete analysis of the condition of your beer as it relates to density, alcohol, original extract, real extract, degree of fermentation, and calories (see table 9.2).

3 "Alcolyzer Beer Analyzing System," Anton Paar (website), accessed September 16, 2018, https://www.anton-paar.com/us-en/products/details/alcolyzer-beer-analyzing-system/.

THE FORCED FERMENTATION TEST

A very important QCP in every brewery is determining if your beer is truly at its final gravity. This is especially important for new brands, when determining the effects of hop creep, or when bottle conditioning. After your fermentor is full of pitched and aerated wort, aseptically (see chapter 14) pull enough sample for at least one SG test. Place the sample on a stir plate or shaker table at 60°F to 80°F for 48 hours or until the fermentation activity stops. The constant stirring, small volume, and warmer temperature accelerate the fermentation process. This allows you to determine the practical limit of attenuation in your main fermentation for this wort with this yeast, referred to as the terminal gravity. Knowing this number will help you make important decisions regarding your beer during fermentation, especially if it is a new product or a recipe change.

To introduce you to the concept in the most approachable way possible, the above method is a modified version of *ASBC Methods of Analysis* "ASBC Wort-5: Yeast Fermentable Extract." If you decide to formalize your lab program, consider moving to the Wort-5 method for increased consistency and reliability.

WHERE TO MEASURE DENSITY

Wort

Measuring the density of the first runnings can give you a rough idea of the potential extract of your grist. This measurement can also help dial in the water-to-grain ratio of your mash. A density measurement of the last runnings can be used as a cutoff point for wort collection. The gravity at which you stop collecting last runnings will be determined by experience and tracking in the brewhouse. Overextraction can pull in silicates, phosphates, polyphenols (tannins), and nitrogenous components of high molecular weight into the wort. The latter two groups are potentially haze-forming materials, and polyphenols can also be astringent and bitter in flavor.

Now we come to a non-negotiable measurement for every brewer: original extract, or OE. Obtain your OE after the boil and before adding yeast. This measurement signifies the starting point of your fermentation and is the basis for your mash efficiency calculations.

Cellar

Measuring the density of your fermenting beer daily is the bedrock for making a consistent product. The measurements confirm yeast health, track fermentation progress, and indicate the fermentation endpoint. Being proactive and tracking fermentation over time will provide invaluable information that can be used for troubleshooting when problems arise. Variation in fermentation performance and/or overattenuation can be indicators of poor yeast health, wort aeration issues, or infection with wild yeast or bacteria. Catching something like this before the beer is packaged can save you from a costly recall, or worse, a tarnished reputation in the eyes of the consumer.

Package

For a distributing brewery, getting a density measurement of beer in packaging is like taking a snapshot of the condition of your beer just before it leaves your hands. If there is ever a consumer/retailer complaint you can reference this measurement and compare it to that of beer in your library (and, hopefully, the returned beer). You can then make an informed decision of your next steps. Even if your beer never leaves your building, I cannot stress enough how important these packaging measurements are for troubleshooting.

SAMPLING PLAN

When drawing up a sampling plan for density measurements, step control is the name of the game. Just as simple as it sounds, a step-control sampling plan is when you take and record measurements at every step of your process. In table 9.3, each QCP represents a step in your process. Recording these values over time will give you a target to aim for that will produce the desired result for the brand.

Identify QCPs from table 9.3 that are relevant to your process and add them to your quality control plan (chap. 6). Record and analyze the data collected using methods that will be covered in part 4 of this book. In part 4 of this text, we will learn how to use statistical process control methods on the collected data for each QCP and set up targets and tolerances. If a specific measurement is found to lie outside of your tolerances, you can use your understanding of the density measurement to speculate why this happened. This is the first step of root cause analysis. In chapter 23 we will learn how to take that further and fix the root cause so it does not happen again using problem solving methods.

TROUBLESHOOTING

The hydrometer is quite possibly the most used instrument across the entire brewery. When there are discrepancies or inconsistencies in hydrometer readings it can leave you scratching your head. Work down this list and try to get to the bottom of the issue.

General (Hydrometer)

- Confirm the working range of your hydrometer, which is typically either 59°F (15°C) or 68°F (20°C) and ensure your sample is at this temperature.
- Verify your hydrometer is calibrated by placing it in distilled water at the temperature appropriate to your hydrometer's calibration (i.e., 59°F or 68°F). The distilled water should read 1.000 SG (0.0°P). Next, prepare a 10% wt./wt. sugar solution, which should read 1.040 SG (10°P).
- If your hydrometer is out of calibration, ensure it is clean, free of dried wort, and not cracked and filling with sample.

If these are not the issues, some may suggest altering the mass up or down to adjust. I find that a bit clumsy and impractical. You are better off ordering a new hydrometer that works correctly out of the box. Make a note if hydrometers from a particular source regularly fail and take steps to find a better supplier.

- Ensure samples are degassed properly
- Check for consistent sampling temperature between users
- Check users read from the same part of the meniscus
- Ensure hydrometer and cylinder are clean and free from soils

Other Density Measurement Instruments

The most common issues with density measurement instruments stem from cleanliness. Wort and beer are very sticky—be sure to clean the instrument after each use according to the manufacturer's recommendations. If you are still having issues, check calibration against a known standard such as distilled water or a sugar solution of known concentration.

TABLE 9.2 DENSITY CONTROL POINTS CHART

QCP	When and how often	What you are looking for
Wort	First runnings	Potential wort extract
Wort	Final runnings	Wort collection cutoff point
Kettle full	Pre-boil	Pre-boil extract (adjust density up or down if needed)
Knockout	Each knockout	Obtain OE
Fermentor	Daily gravity readings	AE reduction, indicators of healthy fermentation
Post-dry hop	Daily	Hop creep*
Conditioning and clarification	Daily	Ensure final attenuation is achieved and maintained
Brite tank	Daily	Final AE to calculate ADF and RDF; log results set standards for the brand; good reference point
From package	Day of packaging	Final AE to calculate ADF and RDF; log results set standards for the brand; good reference point
Library beer	On schedule (weeks, months) over shelf life	Compare to packaged reading; flag and investigate any changes

* Hop creep is increased attenuation following dry hopping.
ADF, apparent degree of fermentation; AE, apparent extract; OE, original extract; RDF, real degree of fermentation

10 TEMPERATURE

Temperature is embedded into the fabric of the brewhouse. Of the big three brewhouse measurements (density, pH, and temperature), temperature may be the most intuitive. Temperature dictates what we decide to wear every day; at an early age we learn not to put our hands on the stove; and there is nothing better than a cold beer on a sweltering summer day. Yet, the concept of temperature can be difficult to define. Ordinarily, temperature is used as a quantitative measure for classifying how matter appears to be hot or cold. So, when you heat (add energy to) the kettle, the temperature of the wort increases. Although related, heat and temperature are not the same thing. To be more specific, matter is made up of moving particles (molecules), with each molecule having its own speed of motion. Kinetic energy is the energy matter possesses because it is in motion. As the movement of an atom or molecule speeds up, its kinetic energy increases; as it slows down, its kinetic energy decreases. Temperature is a physical parameter that is proportional to the average kinetic energy of atoms and molecules, it is not a unit of energy itself. The hotter an object is the more the molecules are moving, and the more the molecules are moving the higher the temperature is.

For a brewery, there are a lot of specific topics to cover when it comes to temperature. To take one example, temperature dictates the condition of raw ingredients before they make it to the brewery—the color and flavor of malt grains is dependent on the kiln temperature and duration used by the maltster. Brewery processes involve heating and cooling large quantities of liquid, so a brewer should have an appreciation and understanding of heat transfer and refrigeration. These topics are covered in detail in other brewing texts. For the QA/QC scientist in a small brewery, it is important to understand the implications of temperature for biochemical reactions throughout the brewing process—specifically enzymatic activity, ester production, diacetyl production and uptake, flocculation, and temperature's effect on shelf life—which is where this text will focus its attention.

Every chemical reaction is dependent on a number of variables. Generally speaking, those variables are time, temperature, concentration, surface area, and the presence of a catalyst. Just glancing over those variables one can see that some are easier to control than others. Practically speaking for brewers, time and temperature are the most easily controlled.

MEASURING TEMPERATURE

Temperature Scales

There are three temperature scales in use today: Fahrenheit, Celsius, and Kelvin. Fahrenheit and Celsius are based on the freezing point and boiling point of water. I am sure you are familiar with both scales, as brewers use them every day. It is helpful to be able to convert temperature (T) between the two using the following equations:

$$T°\text{C} = \frac{5}{9} \times (T°\text{F} - 32)$$

$$T°\text{F} = \frac{9}{5} \times (T°\text{C} + 32)$$

Kelvin is not typically used in day-to-day brewery operations and will not be used in the discussion here.

How to Measure

Glass bulb thermometer. Most people are familiar with glass bulb thermometers. When heated, the alcohol inside the bulb expands and is forced into the narrow vacuum tube that forms the thermometer column. The temperature can be read off the calibrated scale that is printed onto the column.

While there is a place for glass bulb thermometers in the brewery lab, they should never be brought into the production area because they are fragile and pose a physical hazard if broken. This is no-no in the world of GMP.

Bimetallic thermometer. A bimetallic thermometer is made up of two metals with different thermal expansion coefficients that are joined to form a bimetallic strip. The metals expand at different rates with changes in temperature, causing the bimetallic strip to bend a certain way. Using levers and a scale, a thermometer is made. Bimetallic thermometers are common in regulating thermostats and dial thermometers.

Resistance thermometer. Also called a resistance temperature detector (RTD), a resistance thermometer is probably the most common sensor used to measure temperature in the brewery. It responds to changes of resistance in an electrical circuit. You will find resistance thermometers on brew decks controlling kettles, on fermentors and brite tanks that use glycol, and in handheld digital probe thermometers.

Thermocouple. A thermocouple produces a temperature-dependent voltage when subjected to heat. This voltage can be interpreted to measure temperature.

Where to Measure

Mash water. Taking the temperature of the water being added to the grist in the mash tun is one of the first steps you can take to set you up for a successful brew day. Knowing the mash water temperature, weight of the ingredients, ambient temperature of the raw materials, and the speed and volume in which hot water is added means a predictable mash temperature can be achieved.

Mash. In the mash, we convert complex starches into simpler sugars. Enzymes that are activated in malting are reinvigorated in the mash. The activity of enzymes depends above all on temperature. Enzymes are the catalysts in these biochemical reactions and each enzyme works optimally within a specific temperature range (table 10.1). The most important enzymes in the mash are alpha-amylase and beta-amylase. Depending on what fermentable sugar profile you would like in your wort, you can manipulate these enzymes by mash

TABLE 10.1 OPTIMAL TEMPERATURE RANGES FOR MASH ENZYMES

Enzyme	Optimum temperature (°F)	Optimum pH	Inactivation temperature (°F)
Alpha amylase	158–165	5.6–5.8	176
Beta amylase	139–149	5.4–5.5	158
Beta glucanase	104–113	4.5–4.8	131
Peptidase & protease	113–131	4.6–5.3	158
Ferulic acid esterase	100–113	No data	149

Note: Activity can occur outside optimal ranges

temperature. At low temperatures, enzyme activity will be slow at best; once above its optimal temperature, an enzyme will begin to denature rapidly. If you are having erratic fermentations or your real degree of fermentation (RDF) is out of specification, look back at your mash temperature. Wort composition is determined by the action of enzymes, therefore, if you are off by a couple of degrees the mash enzymes will essentially produce a different wort.

Fermentation. Yeast and temperature go hand in hand. The type of yeast you choose will determine your fermentation temperature. Yeast produces many flavor compounds as a consequence of biochemical reactions associated with growth. The flavors a yeast produces are completely strain dependent and will vary based on temperature. In general, a higher fermentation temperature will increase ester production, but this does not come without side-effects. Higher temperatures may also lead to acetaldehyde or higher molecular weight alcohols that taste solventlike and unpleasant. If the temperature is too low, fermentation can slow down and cause the yeast to prematurely flocculate. This may also allow wild yeast or bacteria to gain a foothold in your wort and take over.

Important non-enzymatic reactions are also affected by temperature. These are reactions that take place spontaneously without requiring the activity of yeast enzymes; in other words, these reactions are purely chemical rather than biochemical, although the yeast may be responsible for synthesizing the precursors or breaking down the products. One such example is the diacetyl rest. Diacetyl (a vicinal diketone) is considered an off-flavor compound in beer. While diacetyl itself can be metabolized by yeast, the formation of diacetyl from acetolactate is a chemical reaction that is accelerated at higher temperatures. The purpose of a diacetyl rest is to raise the temperature of the young beer in the fermenting vessel so that the rate of the acetolactate oxidation reaction will increase. Any acetolactate present in the beer oxidizes to diacetyl in a timely fashion, at which point the yeast can take up the diacetyl and metabolize it (see "Diacetyl (VDK) Force Test" in chap. 13).

Finishing/cellaring. Cooling the beer after fermentation is complete allows for flocculation of yeast, proteins, and tannins. Again, the temperature chosen will be strain and process specific. Typically, below 50°F for ales and below 40°F for lagers.

Temperature and CO_2. Henry's law tells us that the amount of a gas dissolved in a liquid is directly proportional to the partial pressure of the gas at a constant temperature. That last part is important—the degree to which gases are soluble in liquids is temperature dependent. This can be seen on a chart of CO_2 solubility against temperature (see fig. 16.1 on p. 159). Note that the solubility of gases decreases as temperature increases. This is due to the higher kinetic energy of a warm liquid, which causes gas molecules to break their intermolecular bonds with the solvent molecules, thus escaping from solution.[1]

Warehousing and storage. Even after the brew, cellar, and packing teams have finished their job successfully (produce a physically and microbiologically stable beer), temperature is still a consideration. As soon as beer is packaged it begins to change, that is, deteriorate. How fast this happens depends on the composition of the beer (hops, alcohol, and carbonation) and its exposure to oxygen throughout brewing, cellaring, and packaging (fig. 10.1).

Time and temperature act together (along with any residual oxygen or other oxidizing agents that might be naturally present in beer) to cause flavor deterioration. Lower temperatures during shipping and warehouse storage will preserve beer quality. In tests, beers kept at 39°F (4°C) were shown to have a shelf life of 112 days. This was reduced to only 28 days when kept at 68°F (20°C). Not every beer ages the same way, however. In-house testing should be done to determine the actual shelf life of each brand. Warmer conditions increase the rate of many chemical reactions, including oxidation reactions that involve oxygen directly or other oxidants such as mineral ions (fig. 10.2). Quite rapid flavor changes are to be expected if consistently low temperatures are not maintained.[2]

---

1 Stephen R. Holle, *A Handbook of Basic Brewing Calculations*, (St. Paul, MN: Master Brewers Association of the Americas, 2003), 65.

2 "Best Practices Guide to Quality Craft Beer," Brewers Association (website), accessed October 20, 2018, https://www.brewersassociation.org/educational-publications/best-practices-guide-to-quality-craft-beer/ (see p. 7–9 of PDF).

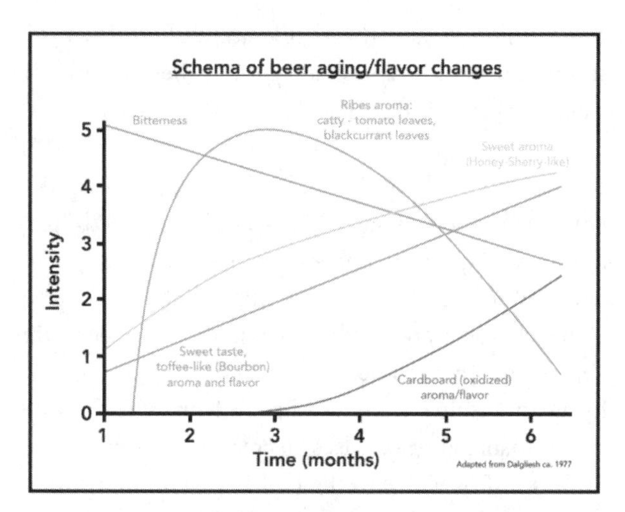

Figure 10.1. Visual representation of flavor attribute development in relation to beer aging time.

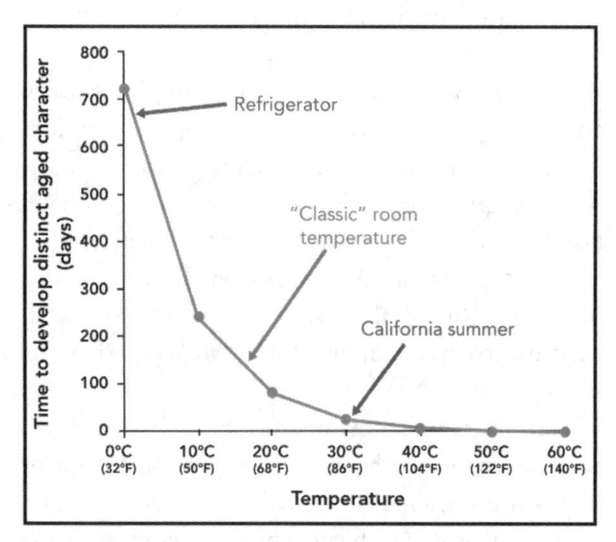

Figure 10.2. Shows the relationship between storage temperature and the time it takes for beer to developing distinct age-related flavor attributes.

SAMPLING PLAN

Controlling temperature is the first step in controlling the rate of reactions happening during your process. Hitting your temperatures, especially during the mash and throughout fermentation, will lead to a consistent final product (table 10.2).

TROUBLESHOOTING

The biggest issue that arises with temperature measurements is calibration. When I discussed the mash earlier, I noted that you should look to your mash temperature if you notice something amiss with your fermentation. If you think you have chosen the appropriate mash temperature, then an RDF out of specification may indicate your measuring equipment is not properly calibrated.

If you have the resources, obtain a thermometer that has been precisely calibrated using the American Society for Testing and Materials (ASTM) methods. Keep it safe in your lab because these are pricey, and use it to check and calibrate the other thermometers around the brewery. An ASTM-calibrated thermometer is convenient but not necessary. Calibration can be checked against reference points on the temperature scale, such as the freezing, boiling, or triple point of water. *Keep in mind these are dependent on your current altitude and barometric pressure.* A relatively precise ice-point bath calibration check can be done as follows:

1. Fill a vacuum insulated bottle with finely shaved ice
2. Add just enough water to fill voids
3. Stir and pack down shaved ice
4. Insert thermometer and allow ample time to equilibrate
5. Calibrate according to manufacturer's instructions or note temperature offset

Note this is a very general overview of the ice-point bath calibration process but should be plenty precise for the craft brewer. For a more detailed process look up the 2011 winter issue of *Cal Lab.*[3]

[3] Jerry L. Eldred, "Preparation and Use of an Ice-Point Bath," Metrology 101, *Cal Lab: The International Journal of Metrology* (Oct/Nov/Dec 2011), 20.

TABLE 10.2 TEMPERATURE CONTROL POINTS CHART

QCP	When and how often	What you are looking for
Strike water	During mash in	Appropriate temperature to achieve desired mash temp
Mash rest	Each mash rest	Targeted enzyme temperature range
Knockout	Throughout	Ensure wort is in desired range for yeast requirements
Fermentation	Daily	Confirm optimal fermentation conditions; confirm glycol is not malfunctioning
Conditioning and clarification	Daily	Monitor conditions
Brite tank	Cold crash for carbonation	Ensure beer is sufficiently cooled for carbonation procedures
Packaging	Throughout	Ensure correct conditions for keg, can, or bottle filling
Beer cooler	Daily	Ensure optimal serving temperature for draft system balance

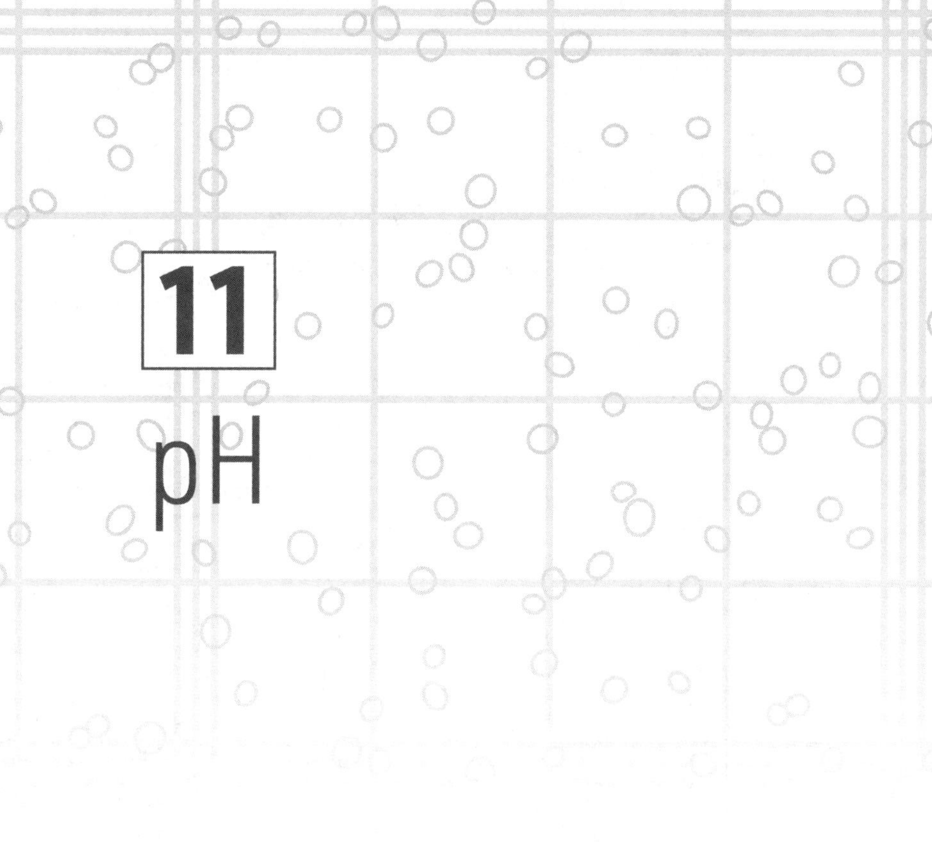

11

pH

A concept first introduced by Soren Sorenson in 1909 while working out of the laboratories at the Carlsberg Brewery, pH has long been discussed in the context of brewing performance and beer quality. Along with density and temperature, pH is one of the most informative measurements used on the brewery floor and in the QA/QC lab. It can be measured with relative ease throughout your process, so understanding, tracking, and controlling pH will be instrumental in achieving a consistent final product.

A SIMPLE EXPLANATION OF pH

pH, which stands for "potential of hydrogen" or "power of hydrogen," is the negative logarithm of hydrogen ion activity in solution. This can be expressed as

$$pH = -\log[H^+].$$

In most solutions, the concentrations of H^+ and OH^- are very small and can cover several orders of magnitude. To make these values less unwieldy, the pH scale uses logarithmic notation (table 11.1). This means that when the pH value drops by 1.0, for example, from 6.0 to 5.0, there is *ten times* more hydrogen ion activity in the pH 5.0 solution than the pH 6.0 solution. Also note that hydrogen activity increases when pH decreases because pH is a negative logarithm.

As shown in table 11.1, in normal use the pH scale runs from zero to 14.0, where 7.0 is neutral because the hydrogen ion (H^+) activity is equal to the hydroxide ion (OH^-) activity. Values below 7.0 are acidic, which means H^+ activity is greater than OH^- activity. Values above 7.0 are basic, which means H^+ activity is less than OH^- activity.

TABLE 11.1 pH SCALE SHOWING HYDROGEN ION ACTIVITY

	pH	H^+ activity in mol/L (M)	
		Standard notation	Scientific notation
Acid	0	1	1×10^0
	1	0.1	1×10^{-1}
	2	0.01	1×10^{-2}
	3	0.001	1×10^{-3}
	4	0.0001	1×10^{-4}
	5	0.00001	1×10^{-5}
	6	0.000001	1×10^{-6}
Neutral	7	0.0000001	1×10^{-7}
Base	8	0.00000001	1×10^{-8}
	9	0.000000001	1×10^{-9}
	10	0.0000000001	1×10^{-10}
	11	0.00000000001	1×10^{-11}
	12	0.000000000001	1×10^{-12}
	13	0.0000000000001	1×10^{-13}
	14	0.00000000000001	1×10^{-14}

pH: A CLOSER LOOK

The pH scale is based on the autoionization of pure water (H_2O). The molecules in pure water can react with each other, whereby a hydrogen ion dissociates from one water molecule, leaving OH^-, and reacts with another water molecule to form H_3O^+ (a hydronium ion).

$$H_2O + H_2O \rightleftharpoons H_3O^+ + OH^-$$

The bidirectional arrow ($\rightleftharpoons$) indicates this reaction is reversible, that is, H_3O^+ can lose its extra hydrogen ion to leave H_2O, and the released H^+ then reacts with a free OH^- to form the second H_2O. This donating (losing) and accepting (gaining) of H^+ constitutes the activity of H^+ in solution. In very dilute solutions, the activity of H^+ is equivalent to its concentration, represented as $[H^+]$. Although strictly speaking it is H_3O^+ and OH^- that is formed, we typically equate this with the activity of H^+ and OH^-. Thus, in water, the pH is expressed as

$$pH = -\log[H_3O^+]$$

but we can say this is functionally the same as

$$pH = -\log[H^+].$$

Note that in pure water H_2O is the only source of H^+ and OH^-. At 77°F (25°C) these two ions exist in equal concentrations in pure water, that is, $[H^+] = [OH^-]$. In this case, it has been experimentally determined that $[H^+]$ is 1×10^{-7} mol/L (i.e., it is an H^+ 1×10^{-7} M solution). Thus, when $[H^+]$ is the same as $[OH^-]$, we have a pH of 7, or neutrality. This is why distilled water is generally considered to be pH neutral, at least at room temperature.

At this point, it is worth noting that when stating a pH value, it is important to know the temperature of the solution. Like any chemical reaction, the reversible autoionization of water is affected by temperature. As the temperature increases above 77°F, the reaction is pushed further to the right and H^+ activity is higher, resulting in a seemingly lower pH value. The reverse happens at lower temperatures, H^+ activity decreases and the resulting pH is seemingly higher. Knowing this relationship, it is good practice to note the temperature of the solution when logging pH measurements. Consequently, measuring and recording pH values at the same temperature every time will provide useful records and enable you to troubleshoot effectively and make adjustments when necessary.

In summary, if there is an excess of H^+ above 1×10^{-7} M then the pH is below 7.0 and the solution is acidic. If there is an excess of OH^- above 1×10^{-7} M then the pH is above 7.0 and the solution is alkaline.[1]

Hydrogen Ion Sources

A reasonable question to ask at this point is where these hydrogen ions come from. There are three main sources of H^+ in brewing. We can see from the example above that water itself is a source of H^+. The other two sources are H^+ from weak acids and the salts of weak acids. A simple definition of a weak acid is a compound that partially dissociates to release H^+ when in aqueous solution (i.e., when dissolved in water). Partially dissociates means that only a small proportion of the weak acid molecules will lose their H^+ in solution. The stronger the acid the more readily its H^+ will dissociate in aqueous solution. The product of a weak acid (HA) dissociating in solution is its conjugate base (A^-) and H^+.

$$\text{weak acid} \rightleftharpoons \text{hydrogen ion} + \text{conjugate base}$$
$$HA \rightleftharpoons H^+ + A^-$$

[1] Charles Bamforth, "pH in Brewing: An Overview," *MBAA TQ* 38, no. 1 (2001): 1–9.

A simple definition of the salt of a weak acid is the product of the neutralization reaction of a weak acid and a strong base. This reaction yields two products: water and an ionic compound that is the salt of the weak acid. For example, sodium hydroxide (NaOH) is a strong base commonly used to neutralize weak acids:

$$\text{weak acid} + \text{strong base} \rightleftharpoons \text{water} + \text{salt}$$
$$HA + NaOH \rightleftharpoons H_2O + NaA$$

Because the salt is an ionic compound it readily dissolves in aqueous solution, with the result being a solution containing the sodium ion (Na^+) and the conjugate base (A^-). The Na^+ does not affect pH. There is a bit more to it than that, but for our purposes it will do. The most important inorganic acids for brewers are carbonic acid (and salts) in brewing water and phosphoric acid (and salts) from malt. Probably the most important organic acid is phytic acid from malt, but many others exist.[2]

Buffers

To understand how to control the pH of your mash, wort, or beer, you need to understand the concept of buffers. Because of its constituents, we can think of wort and beer as a buffer solution. *A buffer is a solution that resists changes in pH when H^+ or OH^- are added to it.* To do this, a buffer solution must contain both a weak acid (HA) and its conjugate base (A^-). As we saw above, a conjugate base is what is left over after an acid has donated a hydrogen ion during a chemical reaction.

$$HA \rightleftharpoons H^+ + A^-$$

By definition, a weak acid is one that only rarely dissociates in water, that is, only rarely will the acid lose its H^+ to water. Likewise, the conjugate base is a weak base, which means it rarely gains H^+ from water. So, [HA] and [A^-] remain stable because the HA and A^- molecules only rarely react with the water. By mixing a weak acid with its salt, both the acid (HA) and base (A^-) components remain present in the solution in relatively high concentrations. The acid and conjugate base may react with one another but when they do so, they simply trade places and [HA] and [A^-] do not change.

How does a buffer solution resist changes in pH? You may have noticed that the dissociation of HA to H^+ and A^- is reversible. When something perturbs the equilibrium of this reaction by reducing or increasing [H^+], the buffer can compensate for this due to the presence of both HA and A^-. Let us suppose a strong acid is added to the solution, increasing the amount of H^+. An increase in [H^+] pushes the dissociation reaction to the left, so that A^- reacts with the excess H^+ to form more HA, thus removing the excess H^+ from solution. Now let us suppose a strong base is added to the solution. This increases the amount of OH^-, which reacts with H^+ to form H_2O and therefore lowers [H^+]. The decrease in [H^+] pushes the dissociation reaction to the right so that more HA dissociates to H^+ and A^-, replacing the H^+ that was lost. In both cases, [H^+] remains the same, thus, the pH remains stable despite the addition of a strong acid or base.[3]

Buffers are not limitless. If an excess of acid or base is added to a buffer, at some point A^- or HA will be depleted. This fact leads us to the concept of buffer capacity. Buffer capacity describes the ability of a solution to neutralize strong acid or base before its pH changes significantly.

$$\text{buffer capacity} = \frac{\text{moles of } H_3O_+ \text{ or } OH_- \text{ added}}{\text{volume of buffer (L)}}$$

Buffer capacity is proportional to the concentrations of weak acid and conjugate base. When brewing, wort and beer act as buffer solutions, so buffer capacity will vary greatly from recipe to recipe. By no means is buffer capacity a QCP that needs to be recorded, but it is a factor to be considered when adjusting mash or wort pH.

MEASURING pH

How to Measure

pH test strips. These test strips are treated with chemical indicators. When wetted with the sample, the strip changes color based on the pH. The color of the test strip is then compared to a color chart that is supplied. Each color corresponds to an approximate pH level. This is the least precise method but will do if you are only trying to determine generally if a solution is acidic or alkaline.

2 A.J. deLange, "Understanding pH and Its Application in Small-Scale Brewing" *More Beer* (blog), July 18, 2013, https://www.morebeer.com/articles/understanding_ph_in_brewing.

3 Chem Collective (website), "Acid-Base Chemistry," accessed on February 17, 2019, http://chemcollective.org/activities/topic_page/5.

pH meter: Pen, handheld, and benchtop pH meters measure the voltage between two electrodes and display the result converted into the corresponding pH value based on the conductive nature of the solution. Pen-style pH meters are great for their small size and portability, but this also can be their downfall as they are often handled roughly in the brewery and are prone to being dropped. They are typically less precise than handheld or benchtop meters and are often more difficult to calibrate. The sky is the limit when it comes to handheld and benchtop meters, but for under $500 you can find the right meter to fit a brewery's needs. A replaceable electrode is a must, as these have a lifespan of 1–2 years even if properly maintained. Look for the option of at least a two-point calibration with an accuracy of 0.01 pH units. I strongly suggest getting a better pH meter than you think you need, as a quality pH meter will pay for itself over time.

Where to Measure

Brewing water. While brewing water chemistry is out of the scope of this section, a brewer will still want to be sure that when they mash in with their treated brewing water, the target mash pH can be achieved. To learn more about how to precisely treat your brewing water, check out Brewers Publications' *Water: A Comprehensive Guide for Brewers* by Palmer and Kaminski. By measuring the pH of your untreated water, you can see if something in your source water has changed or, worse, contaminated your supply water.

Mash. During the mash, a plethora of enzymes break down starches into sugars. Each enzyme has a preferred temperature and pH range. During saccharification, keeping the mash in the "optimal" pH range of 5.2–5.7 will lead to efficient and predictable extract yield and fermentability.[4] Lowering mash pH is typically done by adding $CaCl_2$, $CaSO_4$, a mineral acid (e.g., phosphoric acid), or an organic acid such as lactic acid. It is worth noting here that highly kilned malts contain additional acids that drive mash pH down. In this case, brewers may need to increase the alkalinity of the mash by adding calcium carbonate ($CaCO_3$). It is best to correct any problems at dough-in because proper pH is needed as much for protein and gum digestion as for the lysis of starch.[5]

Lautering or sparging. Brewers monitor pH during sparging because tannin and silicate extraction from malt husks increases substantially if the pH of the runoff is allowed to rise above 6.0. Tannins will taste astringent or bitter in the final product. It has also been reported that grain bed permeability, and hence lauter and sparge flow rates, increase as pH is reduced to within the 5.2 to 5.6 range.[6]

Boil. During the boil, calcium pulled over from the mash will precipitate as calcium phosphate, which acidifies the wort. Additionally, proteins coagulate, and humulone and related compounds isomerize. Even though these reactions are not catalyzed by enzymes, the pH of your boiling wort will play a major role. In the case of proteins, pH directly affects their electric charge. At lower pH levels, proteins have a net positive electric charge, and at higher pH levels they have a net negative electric charge. Thus, favorable protein agglutination properties and trub formation are dependent on keeping wort pH above 5.0.[7] Precipitation at this point of the process is important, since the trub can be easily removed. If it is not removed at this stage, excess protein can lead to clarification issues that persist into the final product, causing haze. Because we would hope the wort to be between pH 5.2 and 5.6 going into the kettle, you should expect it to finish the boil at somewhere between pH 5.1 and 5.4 (this excludes kettle-soured worts).

Fermentation. Yeast are living organisms and their biochemistry only functions well within a certain pH range. If the pH falls outside the yeast's tolerance range, it could inhibit yeast growth or even kill it. Providing a favorable environment (pH, temperature, nutrition) is the ultimate goal of the brewer in order to have healthy yeast. The pH falls during fermentation as a result of the yeast consuming free amino nitrogen (a buffering constituent of wort) and releasing organic acids. Because it normally drops significantly by the end of the exponential growth phase, pH can be used as an indicator of a healthy fermentation.[8] This is good news for the brewer since the pH drop also inhibits the growth of beer-spoiling microorganisms.

---

4 Bamforth, *MBAA TQ* 38: 1–9.
5 deLange, "Understanding pH," 2013.
6 deLange, "Understanding pH."
7 deLange, "Understanding pH."
8 Bamforth, *MBAA TQ* 38: 1–9.

This pH drop will depend on several factors, including the yeast strain chosen for fermentation. In general, the pH of an ale will drop to 3.8–4.2 and lagers in the range 4.2–4.7. After this initial drop, the pH should change very little.[9] If the pH continues to drop, this could signify the presence of acid-producing bacteria. Keeping good records throughout fermentation can alert the brewer to deviations from the norm. For example, a slower than normal drop in pH can indicate underpitching or underoxygenation, both of which can make the wort susceptible to wild yeast or bacterial infection. If a batch is flagged as an abnormal fermentation, you will be faced with some tough decisions, but gathering evidence can push you to make the right decision for your brewery.

Dry hopping. Considering the popularity of dry hopping in massive quantities, it is worth noting the influence this process has on pH. A 2016 study found that, regardless of the beer's starting IBU and pH, dry hopping had a linear impact on pH, with the pH rising by about 0.14 pH units per pound of hop pellets used per barrel of beer.[10] I have observed this phenomenon in dry hopped beers in our own brewery.

Finished beer. If you brew a particular style or brand of beer made from a familiar set of ingredients and fermented with a familiar strain of yeast under your own set of carefully controlled conditions, that beer should have a particular final pH just as it should have a particular terminal specific gravity. If it does not have a consistent final pH, then something is going wrong somewhere along the way. If it does, it gives you a good indication that the process went as desired and that the product will be what you intended. It is important to record the final pH of the beer in package. These records can be used to troubleshoot a problem if it should arise in the market.

Rinse water. Use the pH of rinse water to determine if a tank is adequately rinsed of cleaning chemicals. Collect water from the final rinse to see if it closely matches the source of the rinse water. This will provide a good indication that the vessel has been properly rinsed.

[9] I.C. MacWilliam, "pH in Malting and Brewing," *Journal of the Institute of Brewing* 81, issue 1 (January/February, 1975): 65–70.

[10] John Paul Maye, Robert Smith, and Jeremy Leker, "Humulinone Formation in Hops and Hop Pellets and Its Implications for Dry Hopped Beers," *MBAA TQ* 53, no. 1 (2016): 23-27.

SAMPLING PLAN

We just brushed over at least a semester's worth of chemistry there. With that in mind, it is useful to think of pH as a measure of the chemical reactions taking place in your beer. As stated in chapter 10, every chemical reaction is dependent on a number of variables. Broadly, these variables are time, temperature, concentration, surface area, and the presence of a catalyst. These variables should be considered when identifying QCPs for your sampling plan (table 11.2). In general, when adding a pH measurement to your sampling plan:

- Obtain the sample to be tested at the same time point in your process (e.g., 10 min. into saccharification rest).
- Measure the sample pH at the same temperature; 25°C is a good benchmark standard (remember, as

TABLE 11.2 pH CONTROL POINTS CHART

QCP	Frequency	What you are looking for
Process water	Beginning of the production day	Is water suitable for brewing and cleaning
CIP rinse water	Every final rinse	pH that matches incoming rinse water
Mash	Each mash step	In range for optimal enzymatic activity
Wort	Lautering/sparging	pH below 6.0
Final runnings	Each mash at end of collection	pH below 6.0
Boil pH	At the beginning of boil	Above 5.0
Knockout pH	At knockout (KO)	5.1–5.6 *
Fermentation	Daily, at least	Indicator of healthy fermentation pH dependent on yeast strain
Brite tank	Daily over storage	Stable pH
From package	On packing day	Troubleshooting checkpoint
Library beer	On schedule over shelf life	pH drop from acid producing bacteria

* This pH range is suggested for traditional ale and lager styles produced using traditional production brewery methods. Your KO pH may differ if employing other methods such as mash or kettle souring, extreme brewing, or historical techniques. The point is, monitor your KO pH in order to be able to repeat your process.

the temperature increases so does H^+ activity, resulting in a lower pH reading).
- Consider surface area when adding salts or acid to adjust pH; ensure there is good mixing and contact area.

TROUBLESHOOTING

Calibrate pH Meter

Before addressing changes in a recipe to achieve your desired pH, assessing and calibrating your meter is an absolute necessity. See your pH meter manufacturer's manual for instructions and guidelines.

- How old are your calibration buffers? In general, pH 4 and pH 7 buffers can last 4–8 weeks after opening, and an alkaline pH buffer (e.g., pH 10) will last 1–2 weeks after opening.
- Was the electrode left out to dry? A dry electrode should soak in storage solution for at least 1–2 hours prior to use and should be recalibrated. However, soaking overnight is optimal.
- If a refillable electrode is visibly contaminated, drain the reference electrolyte chamber with a syringe or capillary pipette and refill with fresh electrolyte.
- Some meters are able to display the probe condition automatically after a calibration, but if yours does not have automatic diagnostics you can manually check the slope and offset beforehand if your meter has mV mode. In a pH 7 buffer, the mV offset should be 0 mV but an offset between ±30 mV is acceptable. To determine the slope take the difference in mV values of your two buffers and divide by the difference in pH values, then divide against the theoretical maximum of 59.16 mV/pH unit. A slope between 85% and 105% is a sign the electrode is working properly.

Example: Suppose your pH electrode reads –12 mV in a pH 7 buffer and +175 mV in a pH 4 buffer. To manually check the calibration you determine the slope following the method above:

$$\frac{175\text{ mV} - 12\text{ mV}}{|4-7|\text{ pH units}} = 163\text{ mV/3 pH units}$$

$$= 54.3\text{ mV/pH unit}$$

$$\text{Slope} = \left(\frac{54.3\text{ mV/pH unit}}{59.16\text{ mV/pH unit}}\right) \times 100$$

$$= 91.8\%$$

Recipe Adjustment

If you wish to adjust recipes to compensate for pH, I have included a useful fishbone diagram in appendix E put out by the ASBC. Fishbone diagrams are a problem-solving tool that can be used for root cause analysis. In general, there is a problem statement that is the backbone of the diagram: in this case, it is beer pH. The rib bones reference a major category that effects the backbone. The offshoots from the ribs are subcategories that can be used to influence the process at a specific point and alter the final outcome by adjusting the beer pH up or down. When problem solving, you can also make a fishbone diagram of your own.

12 YEAST

Forget the brewery dog, yeast is a brewer's best friend. Yeast does the heavy lifting, turning brewer's wort into delicious beer. You rely on your yeast to produce great beer every fermentation, so getting to know it will benefit you both overall. There are many books out there that outline yeast's fermentation cycle and nutritional needs. Here I want to talk about getting to know your yeast through measuring its population, growth, density, and viability. The best way to do this for a craft brewer is with a microscope.

Starting from the moment you receive your yeast from a supplier, you should count cell density and viability to make sure it was not damaged (killed) during shipping and you are getting what you pay for. If you are propagating your own pitch, you will want to ensure there is enough yeast to get the job done. Every brewer is after good process control, so why leave your fermentation up to chance? Verify you are pitching the correct number of yeast cells to achieve a predictable fermentation. You can also use pitch rate to control flavor attributes such as esters, vicinal diketones (diacetyl and 2,3-pentanedione), acetylaldehyde, and alcohols. The correct pitch rate will ensure lag time is reasonable and discourage bacterial growth. Finally, after you pitch it is a good idea to confirm the appropriate number of cells made it into the fermentor. Counting cell density as your pitch reproduces and plotting out a growth curve for your chosen yeast strain can help you identify fermentation problems in the future.

A WORD ABOUT YEAST MANAGEMENT

With so many yeast strains available and the large variety of methods for propagation, feeding, and storage of yeast, I will not cover strain-specific yeast management strategies in this book. However, I am presenting many of the tools and methods of analysis that can be used to monitor your house strain. Microscopy and microbiology techniques can provide you with vital metrics you can use no matter how you decide to manage your yeast. I encourage you to measure, experiment with, and really get to know your house strain. Find out what it likes and run with it. If you take care of your yeast it will return the favor by turning your wort into delicious beer time after time.

YEAST MANAGEMENT TOOLBOX

When it comes to yeast there are several parameters to measure that will give a brewer insight into the condition of their yeast. Yeast **cell density** is simply the number of yeast cells per unit volume of slurry. Cell density is useful for calculating and determining yeast pitch rates and tracking cell growth. **Viability** is reported as the percentage of live yeast cells from total yeast cells in a slurry. **Vitality** refers to the activity or metabolism of the yeast cells. Tracking and understanding these parameters is the first step in successfully managing your house yeast.

In this section we will discuss the tools and techniques needed to determine the cell density of a yeast slurry or starter and calculate the amount of yeast you need to pitch. Then we will build on these methods to discuss determining yeast viability and other related applications around the brewery.

As a starting point, the equipment you will need for cell counting and viability staining includes:

- microscope (w/ 400× magnification)
- hemocytometer (Neubauer improved bright-line)
- capped test tubes for serial dilutions (15 mL centrifuge tubes)
- fine-tip pipettes
- lint-free cloth (Kimwipes™)
- viability stain (methylene blue or similar)
- automated cell counter, tally counter, or app

Microscope

To see yeast (or bacteria) in the lab requires a microscope. Microscopes come in all shapes and sizes and with many options. At the bare minimum, you will want a compound microscope capable of 400 to 1,000 times magnification. The microscope will get a lot of use in the QC lab—if taken care of, it can last a very long time, so buy the best microscope you can with your budget. Spring for the binocular type eyepieces, your lab folk will thank you later.

Hemocytometer

A hemocytometer is a square chamber of a specific depth that has been carved into a piece of thick glass. It is used to calculate the density of cells in suspensions when used in conjunction with a microscope. There are some subtleties to consider when using a hemocytometer; I will go over these later in this chapter.

Accessories

To dispense your sample you will need a transfer pipette, 1 mL or smaller. If a dilution is necessary, plastic capped test tubes, for example, 15 mL centrifuge tubes, work well for serial dilutions. If you want to achieve good accuracy, micropipettes are helpful. In a small brewery lab, 10–1,000 μL micropipettes will be used the most often. Delicate wipes, such as Kimwipes or other lint free cloths, are ideal to clean your hemocytometer.

Viability stain

You have some options when it comes to choosing a viability stain. Methylene blue is probably the most widely used in brewery labs so that is what we will focus on here. There are also fluorescent stains, which require a fluorescence microscope capable of detecting the absorption/emission of light at certain wavelengths. Fluorescence is typically used in more advanced automated cell counting devices.

TABLE 12.1 COMMON TYPES OF VIABILITY STAINING REAGENTS

Traditional	Methylene blue
	Crystal violet
	Trypan blue
Fluorescent	Berberine
	Hemisulfate salt
	Magnesium salt of 8-anilino-1-naphthalenesulfonic acid (Mg-ANS)
	Bis(1,3-dibutylbarbituric acid) trimethine oxonol (DiBAC4(3))

Automated Cell Counter

Cell counters use bright-field imaging, fluorescence imaging, and pattern-recognition software to quickly and accurately identify and count individual cells.[1] Automated cell counters are great for labs who do a lot of cell counts or have many operators taking the counts.

MICROSCOPE BASICS

For many brewers it has been years since they handled a microscope. Let us take a minute to review one of the most powerful tools in the brewing lab.

1 "Cellometer X2 Fluorescent Viability Counter," Nexcelom Bioscience (website), accessed January 3, 2019, https://www.nexcelom.com/nexcelom-products/cellometer-fluorescent-viability-cell-counters/cellometer-x2-fluorescent-automated-viability-cell-counter/.

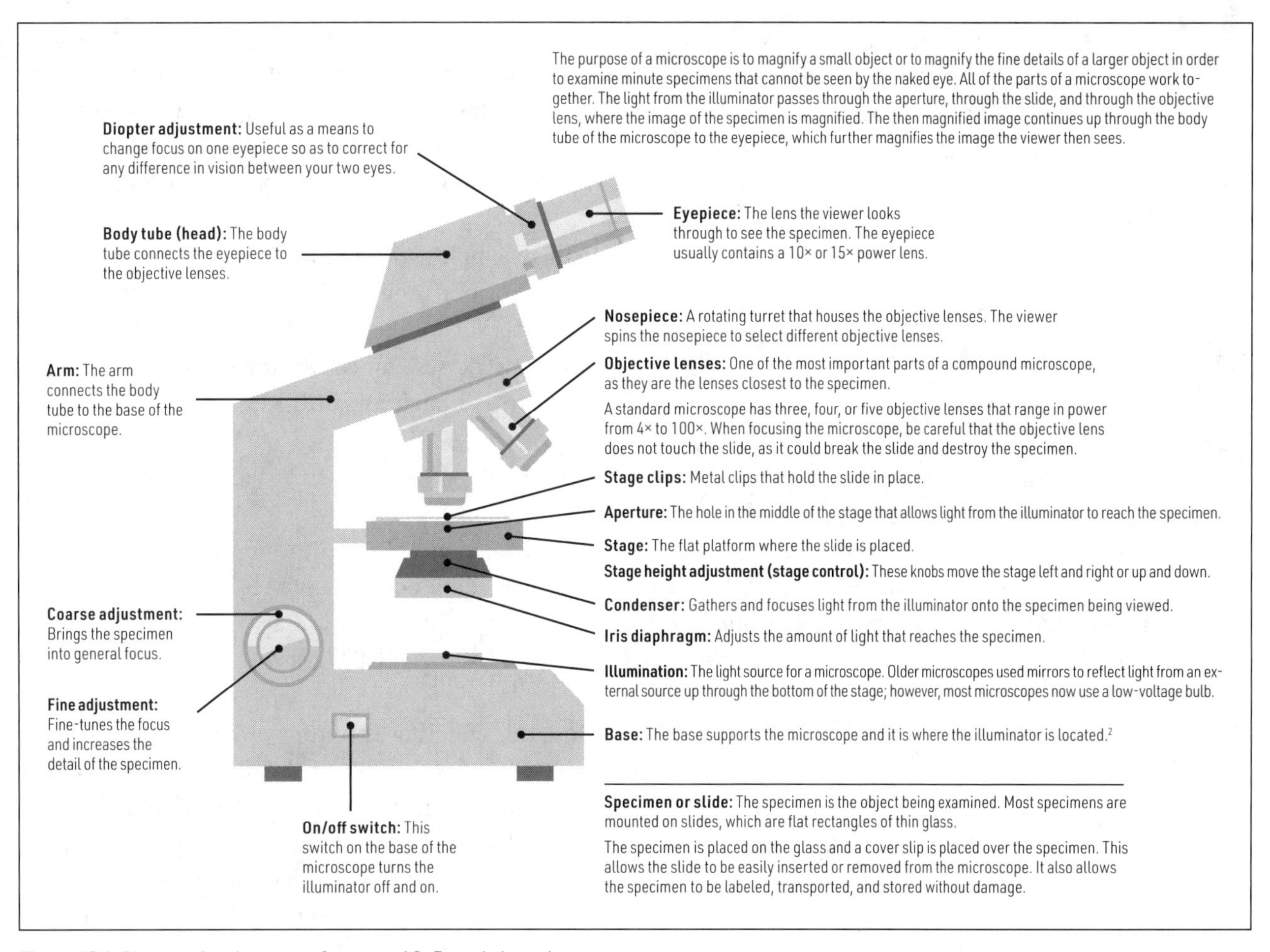

Figure 12.1. Diagram of a microscope. Courtesy of OnFocus Labratories

GETTING TO KNOW YOUR HEMOCYTOMETER

Even with the advanced equipment available to us today, the best way for craft brewers to get to know their yeast is by direct microscopic observation in a hemocytometer. The hemocytometer (Neubauer improved) is a special slide with a precision etched counting grid marked on it, on which a fixed volume of the yeast in solution is placed. By counting the cells in that particular volume under the microscope, the density of cells in the sample can be estimated. We can then extrapolate that information and apply it to our pitch or any other yeast application we desire.

The crux of a hemocytometer is the fixed volume of the counting chamber. Looking closer, we can see the frame of the counting chamber consists of nine large squares, each 1 mm^2 (fig. 12.2). Due to the size of yeast cells, we will be using the large central square, which is divided into 25 medium squares (0.2 mm^2) all bound with triple lines. Each medium square has 16 small squares (0.05 mm^2) inside (fig. 12.3).

Two important things to note:

1. **Only use a hemocytometer coverslip when cell counting. A liquid sample will easily displace a standard glass coverslip thus changing the volume of the counting chamber and leading to erroneous results.**
2. **Even with the proper coverslip, do not overfill the chamber. Overfilling changes the volume of the counting chamber, resulting in errors.**

[2] Hayley Anderson, "Parts of a Compound Microscope," MicroscopeMaster (website), accessed January 3, 2019, https://www.microscopemaster.com/parts-of-a-compound-microscope.html.

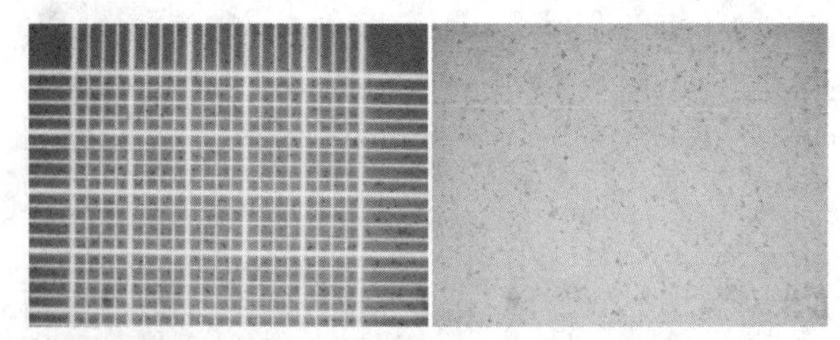

Figure 12.2. View of counting chamber at 10× magnification. Chris Saunders, Escarpment Labs.

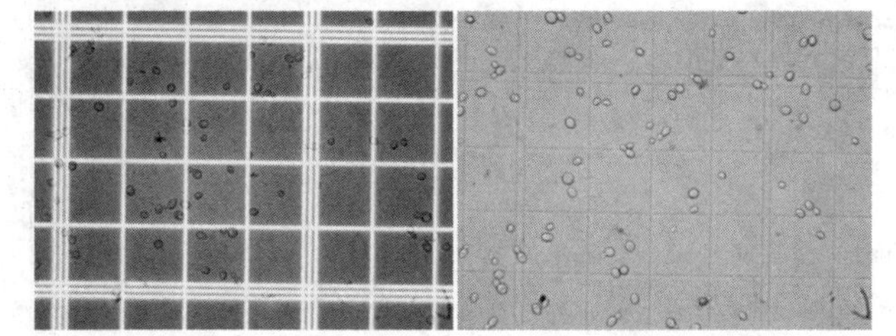

Figure 12.3. View of counting chamber at 40× magnification. Chris Saunders, Escarpment Labs.

When we put the sample under the coverslip, the cell suspension reaches a height of 0.1 mm. Taking these values, we can calculate the volume of one of the large squares:

$$\begin{aligned}\text{volume large square} &= 1\text{ mm} \times 1\text{ mm} \times 0.1\text{ mm}\\ &= 0.1\text{ mm}^3\\ &= 1.0 \times 10^{-4}\text{ mL}\end{aligned}$$

Now that we know the basic dimensions and what our counting area looks like, let us work through an example cell count using brite beer with no dilution needed.

1. Make sure the hemocytometer and cover slip are free of residue by cleaning with a Kimwipe.
2. Place the cover slip over both sides of the hemocytometer counting area.
3. Degas the sample to be counted.
4. Draw up brite beer with yeast in solution into a fine-tip pipette; blot off excess liquid.
5. Carefully fill the counting chamber by touching the tip of the pipette to the edge of the counting chamber/cover slip. Make sure to not overfill the chamber. There should be no overflow into the moat around the counting chamber. There should be no air pockets or CO_2 bubbles under the cover slip, because this would mean the chamber is underfilled.
6. Load the hemocytometer onto the microscope stage. Using the 40× objective lens, find the large central square under the microscope that contains 25 squares bounded by triple lines (fig. 12.3).
7. Starting with the top left square, count all yeast bound inside the triple lines one square at a time, one line at a time, in an organized pattern until all 25 squares are counted. (See rule 1 under "Limitations and Counting Rules" below.)

In this example there are a total of x cells. We know that the size of our properly filled counting area is 1 mm × 1 mm × 0.1 mm = 0.1 mm³. Thanks to the wonders of the metric system, converting cubic millimeters to milliliters is straightforward:

$$0.1\text{ mm}^3 \times \frac{0.001\text{ mL}}{1\text{ mm}^3} = 0.0001\text{ mL} = 1.0 \times 10^{-4}\text{ mL}$$

In our example, we can get the density of cells per milliliter by dividing x cells counted by the volume:

$$\frac{x\text{ cells}}{1.0 \times 10^{-4}\text{ mL}} = x0{,}000\text{ cells/mL} \qquad (12.1)$$

Most cell counting literature presents this as x cells times 10,000 = x0,000 (cells/mL) without explicitly explaining where the factor of ten thousand comes from. Now we must think way back to when we learned about dividing whole numbers by fractions. Recognizing that a number 10^{-4} is less than one, we can rewrite the volume as $^1/_{10000}$ mL. When dividing by a fraction you are actually multiplying by the reciprocal. That give us the equation

$$\frac{x\text{ cells}}{10^{-4}\text{ ml}} = \frac{x\text{ cells}}{\frac{1}{10000}} = \frac{x\text{ cells} \times 10{,}000}{1} = x\text{ cells} \times 10{,}000 = x0{,}000$$

This is important to understand because this equation is the basis of all our cell counts. We can add dilution factors and averages to make our lives easier when counting more dense samples. I will cover these in the next two sections.

SERIAL DILUTION

Before we can obtain a cell and viability count for pitching, we must learn one more skill, the serial dilution. In a thick yeast slurry there are millions of cells concentrated in a small volume. To reduce the density of cells and make it countable using a hemocytometer we must precisely dilute the sample in a series of steps. This can be done by weight or volume. Depending on your lab's resources and capabilities, both are viable options that

DILUTING YEAST SAMPLES

To dilute yeast samples to be viewed under the hemocytometer you will use the serial dilution technique covered in chapter 8 (p. 74) and in the worked example below. The dilution factor you need depends on the cell load in your sample. No dilution is required for most brite or finished beer. A hazy beer may require a 1:10 dilution. Most samples mid-fermentation will require a 1:10 or 1:100 dilution, while a thick yeast slurry will require at least a 1:100 or 1:1,000 dilution. The ultimate goal is to have no more than 50 cells per square evenly distributed across the hemocytometer viewing field.

can produce repeatable results. In the early days of our brewery lab we did not have a volumetric pipette, so we decided to dilute by weight. Even though this is not as accurate as diluting by volume, with the equipment we had it was very repeatable. Diluting by volume is more accurate because very dense slurries are heavier than thin ones. Since the volume is fixed, volume dilutions remove the weight variable and give a more accurate representation of cell density. For most small brewers, simply doing the dilution volumetrically should be sufficient. However, even with training, thick slurries can be hard to work with when measuring volumes, so we will look at examples both by volume and weight.

Find a dilution method that best suits your lab. It is recommended that there be no more than 50 yeast cells in each medium square (0.2 mm^2). For many yeast slurries, a 1:100 dilution will get you there. For more flocculent or dense cultures, a 1:1,000 dilution may be necessary. The more yeast slurry you use, the more representative and accurate the count will be. Many brewers like to complete a serial dilution using 100 mL as the total volume, that is, take 10 mL slurry and dilute with 90 mL distilled water, and mix on a stir plate. Take 10 mL of that first dilution step and add it to 90 mL of distilled water; mix on a stir plate to achieve a 1:100 dilution. In the next example, we will make a 1:200 serial dilution in preparation for a cell and viability count. I am going to assume you have the bare minimum of lab equipment. You will need:

Equipment needed for serial dilution	
By weight	**By volume**
Disposable transfer pipettes Test tubes with caps that can comfortably hold 10 mL of solution; 15 mL centrifuge tubes work well here Digital scale with 0.01 g resolution Distilled water Methylene blue stain 0.01% simple aqueous solution. As your facility expands, you may want to use Fink-Kühles phosphate-buffered methylene blue formulation	Volumetric pipettes or air displacement pipettes Test tubes with caps that can comfortably hold 10 mL of solution; 15 mL centrifuge tubes work well here Distilled water Methylene blue stain 0.01% simple aqueous solution. As your facility expands, you may want to use Fink-Kühles phosphate-buffered methylene blue formulation

For this procedure, obtain a well-mixed sample of slurry. It must be degassed and representative of what is in the brink or propagation tank. Perform a serial dilution of the slurry at a 1:200 concentration.

Serial Dilution Procedure	
Dilution by weight	**Dilution by volume**
1. Weigh out 1 g of slurry in test tube, then add 9 g of distilled water to a final weight of 10 g. Shake well for 2 minutes. This is a 10× dilution (1:10). 2. Weigh out 1 g of the 10× dilution in a test tube, then add 9 g of distilled water for a final weight of 10 g. Shake well for 2 minutes. This is a 100× dilution (1:100). 3. Weigh out 1 g of 100x dilution and add 1 g of methylene blue for a final weight of 2 g. Shake well for 2 minutes. This is a 200× dilution (1:200).	1. Pipette 1 mL of slurry and place it in a test tube, then pipette 9 mL of distilled water to a final volume of 10 mL. Shake well for 2 minutes. This is a 10× dilution (1:10). 2. Pipette 1 mL of the 10× dilution and place it in a test tube, then pipette 9 mL of distilled water to a final volume of 10 mL. Shake well for 2 minutes. This is a 100× dilution (1:100). 3. Pipette 1 mL of 100× dilution, then pipette 1 mL of methylene blue for a final volume of 2 mL. Shake well for 2 minutes. This is a 200× dilution (1:200).

To account for the dilution in the equation for cell density, multiply the cell density in cells/mL that you obtained in equation (12.1) by the unitless dilution factor (in this case, 200). This will give the cell density of the original slurry based on x cells counted in the diluted sample.

Completing the steps above will leave you with a diluted yeast specimen that is ready to be used to fill your hemocytometer chamber for your cell and viability count. However, note that using methylene blue as a diluent is not part of *ASBC Methods of Analysis* "Yeast-3" or "Yeast-4: Microscopic Yeast Cell Counting," but is common practice for the busy craft brewer. When following the ASBC methods, you perform the cell count and viability count separately according to the methods in "Yeast-3" and "Yeast-4."

VIABILITY

Brewers use viability as a way to quantify the condition of their yeast. It refers to the yeast cells that are alive expressed as a percentage of the total cell count. For example, if you count 100 cells in a sample and find 99 live cells and one dead cell you have 99% viability. Again, if you have 100 cells and find 75 live cells and 25 dead cells you have 75% viability. A viability measurement is necessary to ensure you are pitching enough live cells in order reach your desired pitch rate. Low viability can be an indicator of poor nutrition, harvesting, or storage practices.

Methylene Blue

The natural question is, how can you tell if a yeast cell is alive or dead? Microbiologists have developed viability staining methods that provide a visual marker, and by far the most popular stain for craft brewers is methylene blue. Methylene blue will stain dead yeast cells blue while "live" yeast cells remain "colorless" (I use quotations because this is not a hard and fast rule due to some limitations of the technique).

There are three preparations of the methylene blue staining solution recognized in *ASBC Methods of Analysis* "Yeast-3: Yeast Stains." The Fink-Kühles preparation uses a phosphate-buffered solution at about 4.6 pH. If following the international method, the Fink-Kühles preparation is the only way to go, but not everyone has to follow the international method to the letter. You can also get repeatable results using a simple aqueous solution of 0.01% methylene blue, or a 0.01% methylene blue in 2% w/v (if using powdered) sodium citrate dihydrate solution. It is suggested that individual laboratories select the one that best serves their needs (see chapter 8 mixing equation [8.28]). I have had great success using the simple aqueous solution preparation of methylene blue for my daily viability counts.

Limitations and Counting Rules

Methylene blue does have its limitations. The most frequently noted is it may provide misleadingly high viability values as the yeast deteriorates to the point where true viability becomes markedly reduced. Also, methylene blue does not distinguish between "live" cells and their ability to reproduce, meaning it will stain budding cells blue even though they are likely alive. Finally, there can be variations in staining intensity, so some live cells may stain a light blue but should be counted as alive.

Even with these limitations, methylene blue staining can produce very reliable estimations of viability. As with any measurement, you are looking for consistency. Yeast cell and viability counting is definitely a practiced skill and repetition is key. This can get complicated when there are several people performing cell counts in your brewery. By setting up rules, training on them, and following them to the letter, you can make up for some of the method's limitations and variations between users and get consistent results. Here are some rules to get you started:

1. Standardize the counting technique to eliminate the possibility of counting the same area twice.
2. Cells touching or resting on the top and right boundary lines are not counted. Cells touching or resting on the bottom or left boundary lines are counted.
3. A budding cell stained blue is counted as live.
4. A budding cell where the daughter cell is more than half the size of the mother cell is counted as two live cells; when the daughter cell is less than half the mother cell, it is counted as one live cell.
5. Light blue cells are counted as live. This rule presents some challenges between different counters as light and dark blue can be a subjective term—regardless, each counter should make their judgement consistently.

Rules 2 and 3 can be debated because we are assuming budding cells are alive, which may or may not be true. You approach should be to choose to adopt the rules that get you the best results in your fermentations. Once you have chosen these rules, stick to them and make sure the rest of your cell counters do the same. This will help get you the best results possible.

ASBC Yeast-3

Filling the hemocytometer. It is extremely important to fill the hemocytometer correctly because the counting area and correctly positioned cover glass determine the volume over the counting area. The sample is constantly stirred while all 10 counts are performed. A portion of the sample is taken up in a fine-tip pipette, after which the tip of the pipette is wiped dry. After expelling three or four drops of solution, allow a small drop of sample to flow between the hemocytometer and cover slip of both sides of the hemocytometer. The entire counting area must be filled completely, but no part of the sample should extend into the moat. Let the prepared slide stand for a few minutes to allow yeast to settle.

Counting. Counting the yeast cells within the 1-mm^2 ruled area will be accomplished by counting all the cells in the entire ruled area, all 25 of the 0.2-mm^3 squares. A total of 10 counting areas should be counted and averaged to ensure statistical validity. To eliminate the possibility of counting some yeast cells twice, it is necessary to standardize the counting technique (refer to rules 1 and 2 above). Take care with yeast cells that are budding—see rules 3 and 4 above. To obtain an accurate yeast cell count, it is advisable to count no fewer than 75 cells on the entire 1-mm^2 ruled area and no more than about 48 cells in one of the 25 squares. If you are above or below this threshold adjust your dilution to accomodate. Counts from both sides of the slide should agree within 10%. If a dilution is used, the dilution factor must be accounted for in the cell density calculation. Data should be reported to two significant figures.[2]

A WORD ABOUT ASBC YEAST-3

As I mentioned before, yeast cell counting is a practiced skill. I recommend starting with "Yeast-3: Yeast Stains" from *ASBC Methods of Analysis*. The method has gone through scrupulous testing and is validated as the international method. It requires you count 10 hemocytometer slides to produce a statistically significant standard. It will take time, but following this method will build your confidence and ensure you are counting the correct number of cells and reaching your desired pitch rate. This method is not practical for all labs, but once you have mastered this technique you can move on to the quicker craft brewer's method outlined below.

$$x \text{ cells counted} \times \text{chamber volume} \left(\frac{1}{\text{mL}}\right) \times \text{dilution factor} = \text{cells/mL}$$

Remember that you want to express density in cells per milliliter, which involves a factor of ten thousand because the hemocytometer chamber is only 1.0×10^{-4} mL.

$$\frac{\text{cells}}{\text{mL}} = \frac{x \text{ cells}}{1.0 \times 10^{-4} \text{ mL} \times 200}$$

$$= \left(\frac{x \text{ cells}}{\frac{1}{10000} \text{ mL}}\right) \times 200$$

$$= x \text{ cells} \times 10000\left(\frac{1}{\text{mL}}\right) \times 200$$

hence,

$$= x \text{ cells counted} \times \text{chamber volume}\left(\frac{1}{\text{mL}}\right) \times \text{dilution factor}$$

As an example, suppose you have a yeast slurry on which you carried out a 200× serial dilution. Using a hemocytometer, you count 113 cells.

$$\frac{\text{cells}}{\text{mL}} = x \text{ cells} \times \text{chamber volume}\left(\frac{1}{\text{mL}}\right) \times \text{dilution factor}$$

$$= 113 \text{ cells} \times 10{,}000\left(\frac{1}{\text{mL}}\right) \times 200$$

$$= 2.26 \times 10^{8} \frac{\text{cells}}{\text{mL}}$$

3 "A. Dead Yeast Cell Stain (International Method)," in "Yeast-3: Yeast Stains," *ASBC Methods of Analysis*, 8th ed. (St. Paul, MN: American Society of Brewing Chemists, 2011).

Since it is brewing convention to express yeast cell counts in millions, we would say the cell density is 226×10^6 cells/mL, or, more correctly, 226 million cells/mL.

Craft Brewers Quick Method

The procedure for filling the hemocytometer is the same as for ASBC Yeast-3 detailed above. Remember that it is important to fill the hemocytometer correctly to ensure the volume of the sample in the chamber is accurate. While this quick method can save you time on brew day it is less precise than the ASBC method.

Counting. Start with the top left square and count all yeast cells, making note of the number of dead cells, which are stained blue (but remember exceptions in counting rules 3–5 above). Count the top right corner square in the same manner, followed by the bottom right, then bottom left, and finish in the center square of the counting field.

Repeat this procedure for the second counting chamber on the hemocytometer. Counts from both sides of the slide should agree within 10%. If they do not agree, count two more chambers using the same method, then average the four counts and ensure all counts agree with the average within 10%. You can continue until this benchmark is reached or start over and keep a keen eye on your serial dilution, pipetting, or chamber filling techniques.

Example

A yeast slurry was diluted by a factor of 200. You start by counting two chambers and determining cell viability.

Count 1	Total cells =	41	38	42	40	41	Total cells =	202
	Dead cells =	2	0	4	2	0	Total dead =	8
Count 2	Total cells =	38	40	36	42	40	Total cells =	196
	Dead cells =	2	2	0	1	1	Total dead =	6

Determine viability

$$\text{Viability \%} = \frac{\text{total} - \text{total dead}}{\text{total}} \times 100$$

Count 1	Total cells = 202 Total dead = 8	$\text{Viability \%} = \left(\frac{202-8}{202}\right) \times 100$ $= 96\%$
Count 2	Total cells = 196 Total dead = 6	$\text{Viability \%} = \left(\frac{196-6}{196}\right) \times 100$ $= 97\%$

Both counts and viabilities agree within 10% of each other. You now proceed to determine viability of the slurry in viable cells per milliliter.

$$\text{total cells counted} \times 5 \times \text{viability (decimal)} \times \text{dilution factor} \times \text{chamber volume} \left(\frac{1}{\text{mL}}\right) = \frac{\text{viable cells}}{\text{mL}}$$

The factor of 5 comes from the number of squares you counted. This works because there are 25 squares in the chamber. Therefore, multiplying your total cells counted by 5 will produce the average number of cells in the counting area. Plug in your data from above:

Count 1	$202 \text{ cells} \times 5 \times 0.96 \times 200 \times 10{,}000 \left(\frac{1}{\text{mL}}\right) =$ 1.94×10^9 viable cells/mL
Count 2	$196 \text{ cells} \times 5 \times 0.97 \times 200 \times 10{,}000 \left(\frac{1}{\text{mL}}\right) =$ 1.90×10^9 viable cells/mL
Average counts 1 and 2	$\frac{(1.94 \times 10^9) + (1.90 \times 10^9)}{2} = 1.92 \times 10^9$ viable cells/mL

PITCH RATE

Yeast is a major contributor to the flavor and aroma of finished beer. In order to control yeast growth and optimize fermentation flavor, the brewer must choose and achieve a target pitch rate. The term pitch rate refers to the amount of yeast that is added to cooled wort and is typically represented in cells per milliliter. Generally speaking, more cell growth usually results in more flavor compounds. Each yeast strain and each beer in your lineup may have a different optimal pitch rate. While most strains purchased from a yeast supplier have suggested pitch rates these are very general. I suggest you find the optimum pitch rate for your strain and brand on your own. This can be accomplished in just a few simple steps. Choose a pitch rate, count cells and dose at the chosen rate, track fermentation, and note final flavor. There are a few considerations to make when choosing a pitch rate:

- Did you achieve expected flavor result for the brand?
- Was it a complete, timely fermentation?
- How does underpitching affect flavor? (See table 12.2.)
- How does overpitching affect yeast health over the generations?

TABLE 12.2 EFFECTS OF UNDER OR OVERPITCHING YEAST

Underpitching	Overpitching
Increases risk of microbial infection Stuck fermentation Slow to reach terminal gravity Excess diacetyl and sulfur compounds Increased esters Increased fusel or higher order alcohol formation	Reduced esters Extremely fast fermentations (not necessarily a good thing) Autolysis (sharp meaty and sulfur flavor in beer)

After you determine your pitch rate you will need to calculate how much yeast to pitch on brew day. There are several pieces you will need to figure this out: a yeast cell count to determine your slurry's density and viability, your target pitch rate, and what your OE will be on brew day. Once those are known, all you need is the mixing equation from chapter 8 (p. 74, eq. 8.28) to determine how much yeast is needed. Let us take a look at an example.

Pitch rate example

You want to brew a pale ale where:

- batch size = 15 bbl.
- OE = 13°P
- Recommended pitch rate = $\dfrac{1.0 \times 10^6 \text{ cells}}{\text{mL} \cdot {}^\circ\text{P}}$
- Slurry cell count = $1.96 \times 10^9 \dfrac{\text{viable cells}}{\text{mL}}$

The pitch rate above is 1 million cells per milliliter per degree Plato, which is the commonly recommended pitch rate. Calculating how many cells you need per milliliter in a 13°P wort is simple multiplication.

$$\text{Required pitch rate} = \text{pitch rate (cells/mL/°P)} \times \text{gravity of wort (Plato)}$$
$$= 1.0 \times 10^6 \frac{\text{cells}}{\text{mL} \cdot {}^\circ\text{P}} \times 13^\circ\text{P}$$
$$= 1.3 \times 10^7 \text{ cells/mL}$$

So, you need to ensure you pitch your slurry so that you achieve 1.3×10^7 cells/mL wort in the entire 15-barrel batch. Convert barrels to milliliters using a simple conversion factor:

$$15 \text{ bbl.} \times \frac{31 \text{ gal.}}{1 \text{ bbl.}} \times \frac{3785 \text{ mL}}{1 \text{ gal.}} = 1.76 \times 10^6 \text{ mL wort to ferment.}$$

To determine the total number of cells needed, multiply your desired cells/mL by the volume of wort in milliliters that you want to ferment:

$$(1.76 \times 10^6 \text{ mL}) \times (1.3 \times 10^7 \text{ cells/mL}) = 2.28 \times 10^{13} \text{ cells}$$

Now you know you need enough slurry to provide 2.28×10^{13} yeast cells. To determine the volume of yeast slurry to pitch, use the cell count from above

$$\text{volume of slurry} = \frac{\text{total cells needed}}{\frac{\text{viable cells}}{\text{mL}}}$$
$$= \frac{2.28 \times 10^{13} \text{ cells}}{\frac{1.96 \times 10^9 \text{ viable cells}}{\text{mL}}}$$
$$= 2.28 \times 10^{13} \text{ cells} \times \frac{\text{mL}}{1.96 \times 10^9 \text{ viable cells}}$$
$$= 11{,}632.6 \text{ mL}$$

A yeast slurry of around 2×10^9 cells/mL weighs about 1.02 g/mL.[4] Knowing this, you can convert volume of yeast slurry to weight. We can get a good approximation of this by using the scaling equation:

$$11{,}632.6 \text{ mL} \times 1.02 \frac{\text{g}}{\text{mL}} = 11{,}865.3 \text{ g} = 11.87 \text{ kg}$$

If you pitch 11.87 kg of your yeast slurry you will hit your target pitch rate. As you will notice, this method aims to allow you to pitch slurry by weight. If your slurry cell count is much higher or lower than

4 Chris White and Jamil Zainasheff, *Yeast: The Practical Guide to Beer Fermentation*, (Boulder, CO: Brewers Publications, 2010), 125.

2×10^9 cells/mL this will affect the weight of the slurry and you should consider making adjustments when pitching by weight.

If you want to pitch by volume, the approach is simpler. Using the simplified mixing equation derived from (8.28) in chapter 8,

$$C_1V_1 = C_2V_2$$

and taking the same example of a 13°P pale ale wort from above, we have the following:

C_1 = viable cell density in slurry = 1.96×10^9 viable cells/mL

V_1 = volume of slurry to pitch

C_2 = desired pitch rate = 1.3×10^7 cells/mL

V_2 = volume of wort to ferment = 15 bbl.

Plug your parameters into the mixing equation and solve for V_1:

$$\left(\frac{1.96 \times 10^9 \text{ viable cells}}{\text{mL}}\right)V_1 = \left(\frac{1.3 \times 10^7 \text{ cells}}{\text{mL}}\right)(15 \text{ bbl.})$$

$$V_1 = \frac{(1.3 \times 10^7 \text{ cells/mL})(15 \text{ bbl})}{1.96 \times 10^9 \text{ viable cells/mL}}$$

$$V_1 = 0.099 \text{ bbl.}$$

Therefore, 0.099 bbl. of slurry is needed to reach your desired pitch rate. Apply the usual unit conversion equations to convert barrels to gallons or liters of slurry.

$$0.099 \text{ bbl.} \times \frac{31 \text{ gal.}}{1 \text{ bbl.}} = 3.069 \text{ gal.}$$

$$0.099 \text{ bbl.} \times \frac{31 \text{ gal.}}{1 \text{ bbl.}} \times \frac{3.785 \text{ L}}{1 \text{ gal.}} = 11.62 \text{ L}$$

WHERE TO MEASURE

Each of the measurements discussed here will give you insight into how your yeast behaves. By tracking and recording these QCPs, you are essentially listening to what your yeast needs to provide you with delicious beer continuously. You can spot positive and negative trends, putting you in a position to save time, money, and beer.

Yeast pitching. Use the methods described in the previous section to quantify cell density and viability in your slurry, and this will enable you to achieve your desired pitch rate.

Pitch confirmation. It is important to confirm that your calculated pitch rate actually resulted in the desired level of yeast in your wort. Take a post-knockout cell count and also record cell viability. This step can be a real lifesaver—it puts you in a proactive position if a tank somehow gets vastly underpitched or maybe even not pitched at all! Things can happen quickly at this point (e.g., physical settling in tank or some growth from first knockout), so standardizing the time point at which you measure after knockout is recommended.

Yeast growth. Get an idea of how your yeast is growing and settling throughout its time in the fermentor. Since each yeast strain is different, measuring yeast growth here is all about gathering baseline data. Check cell density at set intervals between 12 and 24 hours after pitching yeast. After several fermentations under the same conditions (recipe, pitch rate, temperature, oxygenation rate) you should be able to recognize parameters follow the same pattern, for example, in your pale ale after 24 hours at a pitch rate of 7.5×10^5 cells/mL/°P yeast strain X has doubled twice and its cell density is 3–5 times the original pitch rate. This can serve as an indicator of how the rest of fermentation will go under normal conditions. On the other end, if you consistently see low growth and sluggish fermentations, you can refer to the data, start a plan-do-study-act (PDSA) cycle and determine if you would benefit from more oxygenation, adding zinc, order or propagate a new culture, etc. (More on PDSA in part 4 of this book.)

Post-fining/pre-transfer. Check cell density per milliliter to confirm your fining rate is optimized. Set a benchmark and check cell density as a QCP before transferring to the brite tank. Cell counts above 5.0×10^5 cells/mL can appear as haze in a pale colored beer, you may also notice a sensory difference with elevated cell counts in finished beer. Elevated cell counts at this point will typically be perceived as an unpleasant biting bitterness and a sandy or chalky mouthfeel.

Brite beer and packaged beer. If you are meticulous about clarity you can follow your cell density from the brite tank to the package to ensure there was no carry over from the fermentor or settling in the brite tank. The starting cell count on packaging day can be used as a QCP to determine if the beer at the end of a packaging run is within the acceptable range for that brand. This range is set by you and should be close to the starting cell count so as to achieve consistency in appearance and flavor.

Not all yeast strains behave the same. Use the microscope along with other measurements (e.g., density, pH) to get to know your yeast strain. Pay attention and you will find the right balance between flavor and performance. Not only will your production schedule be happy but so will your taste buds.

SAMPLING PLAN

TABLE 12.3 YEAST CONTROL POINTS CHART

QCP	When and how often	What to measure
Propagation	Each propagation step	Cell density Viability
Pitching	Every brew	Cell density Viability Volume/ weight pitched Track lot and generation
Growth	At least at 24 and 48 hours (more frequently if possible) and at turn down or harvest. Especially important when establishing new brands	Cell density Viability Track yeast growth and flocculation
Pre-transfer to filter, centrifuge, or BBT	Determine your target cell density at each of these stages	Cell density
From package	Day of packaging Over shelf life	Cell density
In storage brink	Before next pitch At a frequent interval to get a sense of how your strain behaves in storage	Cell density Viability

TROUBLESHOOTING

Some of the largest discrepancies I have seen revolve around proper pipetting technique and over/underfilling the hemocytometer. For proper pipetting technique, check appendix E for tips and examples.

For hemocytometer technique, pay careful attention when filling a hemocytometer. There must be no air bubbles or excess liquid in the moats. It is also possible to knock the coverslip off when placing it on the microscope stage; check for movement visually before counting. If slippage has occurred, clean and load the chambers again.

Using methylene blue has its challenges. Do your best to keep your solution fresh and within its buffer specification. Ensure the staining solution has 2–5 minutes of contact time with the slurry.

13
SENSORY ANALYSIS

Whether you know it or not, sensory analysis is probably what sent you down the craft beer rabbit hole. We all have that one beer that was our "Ah-ha!" moment. I still remember mine vividly. It was an Old Rasputin, North Coast's Russian imperial stout. Everything about it was pleasing to me. The jet-black color, the inch and a half of brown-tinted foam on top, rich and complex coffee, chocolate, and raisin flavors, the warming alcohol sensation, all the way to the roast flavors in the aftertaste. I wanted more. Not only more Old Rasputin, but more craft beer in general.

Sensory analysis is an important part of your QA/QC program and often overlooked by the craft brewer. Lack of time and small panel size are the two most cited excuses for not doing sensory at a craft brewery. Initially it will take some time to get organized, but the condition of your beer depends on it. Sensory analysis can be used to confirm or validate the processes that you measure. I do not care if you have a panel of two, you can learn something about your process through sensory analysis (e.g., maybe you are sensitive to a certain attribute, such as diacetyl, but your co-panelist is not). You can avoid a lot of troubleshooting through a rigorous sensory program. Take the time for sensory analysis!

The entire brewing process begins and ends with sensory analysis—we think about the sensory qualities of beer even before brew day. When developing a recipe, we are deliberate over what specialty malts to add to the grist for a desired outcome; after that we do the same thing with the hops and yeast. When ingredients arrive at the brewery we inspect them for quality and flavor before the bag even makes it to brew day. When we take samples throughout fermentation for gravity and pH we should also taste that sample to see if the beer is progressing as expected. Once completed in the brite tank we give the beer a final taste for go/no-go. We taste the beer on draft, and we evaluate it in the can or bottle on day one and intermittently throughout the beer's shelf life. One can argue the most important QA/QC instrument is the tasting glass.

WHAT IS SENSORY ANALYSIS?

Sensory analysis is different from most of the measurements covered in this book. We are using the human senses to try and quantify the levels of thousands of different flavor molecules contained in the beer. Thousands of years of evolution have honed the human palate into a fine-tuned measuring instrument. Our survival depended on it. But, just like each person has a unique personality, each taster has a unique palate. This can make sensory analysis data difficult to interpret. Large-scale sensory programs combat this by thorough training, statistical analysis, and several testing methods to filter out noise in the data. Even if those things are not available to you, that is OK, you can still implement a useful sensory program (no matter your brewery size) and include more sophisticated methods as you grow.

Sensory analysis is using human senses—sight, smell, touch, taste, and sound—to evaluate a consumer product. In our case, the consumer product is beer and we use appearance, aroma, flavor, and mouthfeel to describe it. Sensory analysis may be the most powerful tool in your quality system. When all your measurement equipment breaks down you will still have a glass. When all is said and done, what matters is how the beer tastes in the glass. Do not skip your sensory program, it is as important as your hydrometer.

SENSORY PROGRAM

A sensory program is the system in which you evaluate your beer using your senses. Just like your QA/QC manual, you should write it down. The written sensory program can be as simple as a tasting checklist with comments, or as complicated as a fully fleshed plan with descriptions of what was measured, goals, schedules, methods of analysis, and document control.

What to Measure

Before you start planning your sensory program, let us talk about the basics of what you are measuring.

Appearance. Appearance is, simply, what the sample looks like. You should mention the color; describe the clarity in terms of the haze or glow, opacity or cloudiness, and particulates in suspension. Foam should also be mentioned; is there any; what does the foam color, density, lacing, and retention look like? You could even mention carbonation here, such as bubble size or effervescence.

Flavor. Flavor drives the craft brewing industry. It describes the overall character of the beer, it is what the consumer will most likely describe and what your customers will judge you on. Flavor describes the combination of basic tastes, aromas, mouthfeel/texture, and aftertaste. Splitting flavor into distinct parts can be used to track the condition and trajectory of your beer and give you a better understanding of your process. Let us take a look at each one.

Aroma: thousands of molecules combine in thousands of combinations and provide wonderful aromas for us to smell in our beer. Aromatics are detected in the olfactory region (the nose) perceived through the stimulation of olfactory receptor cells and the free nerve endings of the trigeminal nerve.[1] We perceive aromas through two different mechanisms. First, what we perceive with our nose is orthonasal olfaction. What we smell after we take a sip and subsequently swallow is retronasal olfaction.

Basic tastes: the sensations perceived through stimulation of the gustatory receptor cells enclosed within the taste buds of the tongue. There are five commonly recognized basic tastes: sweet, sour, salty, bitter, and umami (while others exist, such as oleogustus, a.k.a. fat taste, they will not be covered here). The rest of the "flavors" that you taste are contributed by aromas detected by your olfactory senses. You can separate some of these basic tastes by experimenting with a cup of coffee. Plug your nose while sipping cool, black coffee; you should taste only bitter and sour at first. Release you nose, and all sorts of roast, chocolate, and fruit notes come rushing in.

An illuminating exercise that I picked up from Roy Descroters of the Tufts University Sensory and Science Center is to prepare simple solutions of water and one of sweet, sour, salty, and bitter flavor compounds at weak and strong levels (table 13.1). If you taste these blind and in succession you can start to hone your basic taste skills. Go DIY sensory!

Mouthfeel/texture: I lump mouthfeel and texture together here, but they are decidedly different attributes. Mouthfeel is the chemical or physical sensations that are detected in the mouth, nose, and throat. Texture is the consistency or physical description of the sample. Think about eating a fresh jalapeño: you bite in and chew, it is probably crunchy or crisp, which are descriptions of the

1 Liz Pratt, "Introduction to Sensory Analysis and Beer Flavor," (lecture, 2018 MBAA Brewing and Malting Science Course, University of Wisconsin, Madison, WI, November 1, 2018).

TABLE 13.1 SIMPLE DIY SOLUTIONS TO HONE BASIC TASTE SKILLS

Taste	Intensity	Concentration	Example products
Sweet	Slight	5% w/v sucrose	Peanut butter Unsweetened juices
	Moderate	10% w/v sucrose	Soft drinks Vanilla ice cream
	Strong	15% w/v sucrose	Jellies Preserves
Sour	Slight	0.05% w/v citric acid	Milk chocolate Coffee
	Moderate	0.10% w/v citric acid	Soft drinks Ketchup
	Strong	0.20% w/v citric acid	Lemon juice Vinegar
Salty	Slight	0.4% w/v sodium chloride	White bread Canned peas
	Moderate	0.7% w/v sodium chloride	Canned soups Sardines
	Strong	1.0% w/v sodium chloride	Soy sauce Anchovies
Bitter*	Slight	0.05% w/v caffeine	Whole peanuts Milk chocolate
	Moderate	0.10% w/v caffeine	Beer Strong coffee
	Strong	0.20% w/v caffeine	Baking chocolate Quinine water

* Caffeine can be replaced with quinine (tonic water) or iso-alpha acids for those who are caffeine sensitive.

texture; it is followed by a burning sensation, which describes the mouthfeel.

Aftertaste: flavors that remain (or develop) at least one minute after initial sampling. Like the wise Axel Rose once said, "Need a little patience, yeahhhheee yeahhh." Wait at least two minutes and see what happens and use your words to describe it.

Now, I know this is a lot to take in, but you will be glad you know it when your sensory program takes off and you are able to troubleshoot issues that arise. For more training tips and information, see the *Craft Brewers Guide to Building a Sensory Panel* available from https://www.brewersassociation.org/.

Practice

Tasting is a practiced skill, not everyone will be able to clearly articulate what they are tasting. Establishing a common vocabulary based on the criteria above will be step one. We have several tools to help us with this task. First is the Beer Flavor Wheel, which was developed in 1979 by Dr. M. Meilgaard with the MBAA, ASBC, and European Brewery Convention (EBC) in an attempt to standardize the vocabulary necessary for describing beer. The Beer Flavor Wheel is now available for free online (http://www.beerflavorwheel.com). The inner ring lists 14 classes of flavor components. As an example, Class 1 is "Aromatic, Fragrant, Fruity, Floral." The inner tier then represents the more specific flavors that appear within that class, which for Class 1 is broken down into the descriptors alcoholic, solvent-like, estery, fruity, acetaldehyde, floral, and hoppy. The class and inner-tier terms contain common terminology familiar to most people.[2]

[2] Gary Spedding, "The Oxford Companion to Beer definition of flavor wheel," accessed February 21, 2019, https://beerandbrewing.com/dictionary/9ZFDlvjDyv/.

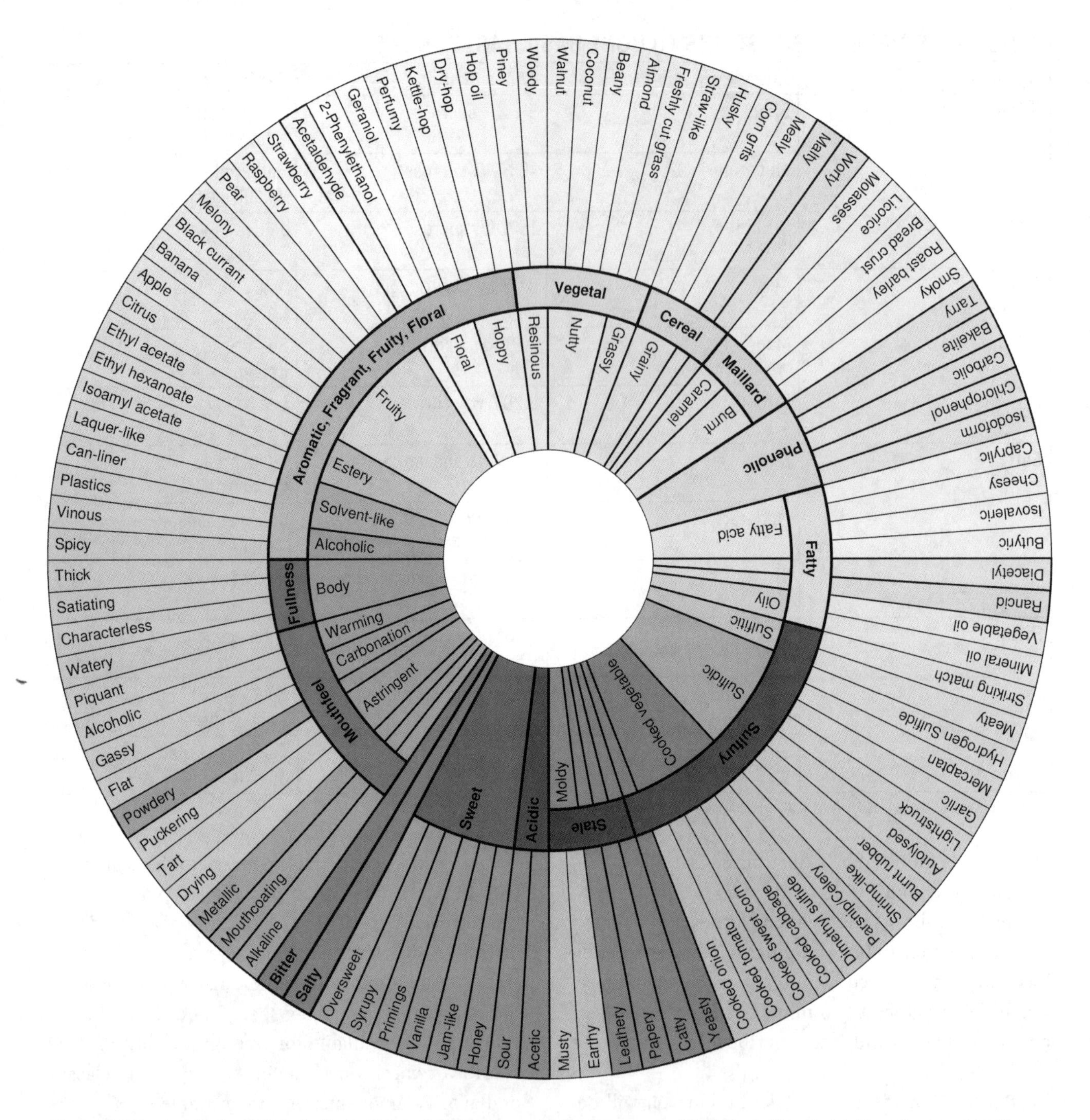

Figure 13.1. Developed in the 1970s by Dr. Morten Meilgaard, the beer flavor wheel has become one of the tools used to articulate flavor. Courtesy of beerflavorwheel.com and Abraham Kabakoff.

The flavor wheel was used as the standard for years, but a lot has changed since then. Beer styles and brewing have evolved dramatically and so has the sensory science. Our second tool is similar to the flavor wheel and is called the Beer Flavor Map. Lindsay Barr and Dr. Nicole Garneau developed and released the flavor map in 2016. The Beer Flavor Map builds on the flavor wheel, it incorporates 40 plus years of research and it is both scientifically accurate and user friendly. Printed copies of the Beer Flavor Map can be purchased at https://www.asbcnet.org/. It is also paired with the Draught Lab® software that can help with tracking and analysis of sensory results.

With a flavor guide of your choosing in hand, gather your tasting panel, pour them a beer (any beer will do), and take 10–20 quiet minutes and write down notes about the modalities mentioned above (appearance,

Name:

Style ABV

Color and Appearence
☐ Appropriate ☐ Not Appropriate ☐ Dark ☐ Light
Comments:

Aroma
☐ Appropriate ☐ Not Appropriate
Comments:

Flavor and Aftertaste
☐ Appropriate ☐ Not Appropriate
Comments:

Mouthfeel/Body
☐Very Light ☐ Light ☐ Medium ☐ Full
Comments:

Carbonation
☐ Appropriate ☐ High ☐ Low
Comments:

Bitterness
☐ Appropriate ☐ Not Appropriate
Comments:

Alcohol
☐ Appropriate ☐ Not Appropriate
Comments:

What did you like best/least about this sample?
☐ Excellent ☐ Very Good ☐ Good
☐ Acceptable ☐ Needs Improvement
Comments:

Name:

Style ABV

Color and Appearence
☐ Appropriate ☐ Not Appropriate ☐ Dark ☐ Light
Comments:

Aroma
☐ Appropriate ☐ Not Appropriate
Comments:

Flavor and Aftertaste
☐ Appropriate ☐ Not Appropriate
Comments:

Mouthfeel/Body
☐Very Light ☐ Light ☐ Medium ☐ Full
Comments:

Carbonation
☐ Appropriate ☐ High ☐ Low
Comments:

Bitterness
☐ Appropriate ☐ Not Appropriate
Comments:

Alcohol
☐ Appropriate ☐ Not Appropriate
Comments:

What did you like best/least about this sample?
☐ Excellent ☐ Very Good ☐ Good
☐ Acceptable ☐ Needs Improvement
Comments:

Figure 13.2. Example form for a description test panel. Based on Beer Judge Certification Program score sheet.

aroma, taste, mouthfeel/texture, aftertaste) using a form that describes each (fig. 13.2). After everyone is done, discuss what you wrote. And there you go, you just held your first sensory analysis session. This is called a description test and it forms the basis for development of true-to-brand descriptions of your beer. Practice like this, as often is convenient for you, but no less than once a week. Once you and your panel feel comfortable, you can take it to the next level.

TASTING IN-PROCESS BEER

Tasting your beer at every stage of the process is extremely important. Using sensory analysis as a form of measurement, you will get to know how your beer is progressing over time. Tasting in-process beer is a different experience from tasting the finished product. If you decide to include in-process beer into your tasting session, first ask yourself what question you are trying to answer, for example, does this beer smell like sulfur? Next, make sure your panel is equipped with the necessary training to answer your question. Finally, the correct testing method needs to be determined to get an unbiased answer to your original question. This can be a difficult task but with a well-trained panel and thoughtful sensory administrator many questions about your beer can be answered.

Developing Brand Recognition, True-to-Brand

Let me ask a couple of questions: what do you *want* your best-selling beer to taste like, and what *does* your best-selling beer taste like? Even though we are talking about the same best-selling beer, the two answers can be vastly different. If you choose to do one thing with your sensory program, I highly recommend training for brand recognition, a.k.a. true-to-brand training. Before you start training you must establish what true-to-brand for that beer is. You can approach this several different ways.

Recipe. You can let recipe influence tasting descriptors. For example, you probably expected orange and citrus notes from all the Amarillo hops you used in your IPA. That may be a leading aroma descriptor for that brand. This approach can work for one-off brands and falls under what you want your beer to taste like, but be ready to be surprised from time to time.

Experience. In my opinion, experience is the best guide to the writing of true-to-brand descriptors. It can take some time. Conduct three to five blind practice sessions on subsequent batches of beer, compile the most frequently used descriptors, and hash out a true-to-brand description based on those. This harkens back to my second question—if your practiced tasting panel tells you that your IPA tastes like geranium, it probably does.

Armed with specific tasting descriptors (flavor attributes) derived from the methods above you can go back and fill out a true-to-brand sheet using appearance, aroma, taste, mouthfeel, and aftertaste. True-to-brand is the foundation of your sensory panel, without it you cannot determine if there are any flaws or deviations. A word of advice when setting up true-to-brand for a particular product: it is normal to have variations in any production process. Be only as specific in tasting descriptors as your process allows. Pick only the dominant characteristic that describes the attribute you are trying to define.

Brewery X Brown Ale - True-to-Brand
Appearance: Clear dark brown; dense white foam that lingers
Aroma: Bread dough, toast
Taste: Moderately sweet
Mouthfeel: Medium body, mildly warming

Looking at the brown ale true-to-brand description you can see that bread dough is a prominent descriptor for the aroma. Suppose you have this true-to-brand attribute in place and your panel notices a heavy roast character is starting to overpower it, this may be a sign that the beer's profile is drifting unwittingly. This is a case where sensory analysis can initiate an investigation. Is the beer still within its true-to-band specification? Is a root cause analysis of the roast character source warranted? Questions only you can answer, but it is better to have your sensory panel identify this drift in the brand before your customer base notices and complains or just stops drinking it.

Formal Production Release Protocol

Establishing true-to-brand is the first step in writing your own production release protocol. An important QCP for any size brewery, production release uses your sensory panel (of two or more) as the check to determine if a beer is ready be moved on to the next step. It is usually performed before the beer goes from brite tank into package, a step typically determined by whether the beer is true-to-brand or not true-to-brand ("TTB or not TTB"). A well-trained panel that understands each true-to-brand attribute is the key component needed in order to write your production release protocol. This is where your sensory program should begin to get formal.

Sensory space. Do not wait to build out a dedicated space for sensory analysis, use the space you have and optimize it to be quiet and free of distraction. Shut off the music, discourage talking, do not allow food or outside drink. Close the doors and windows and try to be away from any production smells that could be distracting (e.g., boiling wort, hops). This space could be your tasting room bar or simply a folding table pulled out of storage. If possible, spread everyone out or construct simple dividers fashioned out of poster board, recycled cardboard, or even slip sheets.

Scheduling. I cannot stress this enough—make the time for sensory, even if it is only one hour a week. I have found success scheduling a formal sensory session at the beginning of every staff meeting. This optimizes panelist turnout, giving more data points to base decisions on. Every brewery is different, so find what works for you, put it in the calendar, and execute. Do not skip a session; just like the gym, if you skip once you may never go back.

Panelists. Invite everyone to be panelists! You never know who your best taster will be. A person with loads of brewery experience does not necessarily mean that they are a better taster. In fact, they may have ingrained bad habits or misconceptions. Have realistic expectations, for instance, panelists will not always agree with each other. Make it clear to the panel that the decision to pack, blend, or destroy a beer does not rest with the opinion of one taster. All panelists should drink critically and make clear comments so that the entirety of the panel will be a data point used in the decision-making process.

Training panelists will take time, so repetition is key. Encourage responsible tasting at every step of the process that they are involved in. Brewers should be evaluating raw ingredients at multiple locations, cellar folk should evaluate in-process beers while taking gravities, packaging should be tasting in brite tanks before a single keg, can, or bottle gets filled. Tasting room employees should taste every keg changed before serving it to a guest. Repetition in this fashion can make you alert to deviations from true-to-brand early on.

Data Collection Forms

There are many different tests that can be presented to a sensory panel. Let us look at a few examples and their uses.

Description testing. A description test is where you present each panelist with a sample and instruct them to systematically evaluate its attributes. This form is great for gathering true-to-brand data, general vocabulary practice, and tracking long-term brand deviation. If followed by a discussion in a class format, the exercise can also be used as a training tool. Little formal training is required to get useful data from panelists.

True-to-brand evaluation. In a true-to-brand evaluation, you present each panelist with a sample. Panelists are instructed to systematically evaluate attributes by marking yes or no for each attribute being present, and also an overall marking section is used for a final true-to-brand decision. If a panelist marks no for true-to-brand, they must qualify the reason why, the more specific the better. This form requires true-to-brand training for data to be valid.

Go/no-go. For go/no-go, present a panel of at least two members with a sample with instructions to systematically evaluate its attributes. Each panelist will mark "go" or "no-go" for each attribute. This form is useful for mid-production samples, such as sending a product from fermentor to the filter or brite tank, or it can be used as a final check to send a beer to package. It can be included easily on production paperwork (fig. 13.3). Be aware, this type of test is only useful with a well-trained panel, but I like it because it can force your panel to raise QC flags that can then be handled before your beer gets into package and out the door.

If you are trained up on true-to-brand but are looking for more information, look to the *Craft Brewers Guide to Building a Sensory Panel* (https://www.brewersassociation.org/).

Color GO_____ NOGO_____	Haze GO_____ NOGO_____	Aromas GO_____ NOGO_____	Carbonation GO______ NOGO_____	Initials
Color GO_____ NOGO_____	Haze GO_____ NOGO_____	Aromas GO_____ NOGO_____	Carbonation GO______ NOGO_____	Initials
Color GO_____ NOGO_____	Haze GO_____ NOGO_____	Aromas GO_____ NOGO_____	Carbonation GO______ NOGO_____	Initials

Figure 13.3. Example of form used to release finished beer from brite tank to packaging line. This is not intended to replace a production release true-to-brand panel.

Data Collection and Monitoring

Holding sensory panel sessions and monitoring the data that come out of them goes hand in hand. At the beginning, data collection can be as simple as compiling flavor attributes brought up in a session and tallying the number of times they come up. You can use data collected in this way to develop true-to-brand. It is a good idea to keep individual responses from each session either in a binder or digital database. This can be useful for recipe development in the future when that beer is brewed again.

For questions with binary responses (e.g., yes/no, pass/fail) statistical analysis can be employed. This can be as easy as a spreadsheet with the tests performed in the rows and a column for each panelists name. When you start to track data like this you can really get to know your panelists, their strengths, and their blind spots. If you know a panelist well, it gives you the ability to get a second opinion on a flagged attribute when you know that taster can detect it in low levels. You can also see if there is a disagreement between panelists. This can present an opportunity for further training, which is important if you want to get consistent and reliable data from your sensory panel.

For larger panels, p-charts are used to monitor the proportion of nonconforming units (i.e., those not true-to-brand). For this kind of analysis, control limits must be set, which going forward are then used to monitor the proportion nonconforming for the process. When a point is outside these established control limits it indicates that the proportion nonconforming for the process is out-of-control. When a process is deemed out-of-control, decisions must be made whether it is a product hold, blend, rebrand, or destroy. Such incidences should also initiate a root cause investigation of some kind.

Get to know your panelists, keep track of what they can identify and even what they misidentify. These little details can help you nail down a process issue in the future. No matter what kind of data collection you decide to do for your brewery, regular and frequent training and practice is paramount to your panel's success.

FLAVOR STANDARDS

Using flavor standards is a terrific way to advance your sensory panel's ability to the next level. A flavor standard is a precisely measured amount of molecular compound that represents an individual attribute or characteristic. It is sort of like having a mixer on your stereo and being able to turn way up your favorite supporting vocal harmony from the background. When training, you can increase the sophistication of your panel by familiarizing panelists with the appropriate chemical names, for example, instead of using the descriptor "buttery" your panelists can identify diacetyl (2,3-butanedione).

There are many companies that produce flavor standards. The most common sources for craft brewers are Aroxa, FlavorActiv, and Siebel. All make very good products that are easy to use, so it comes down to your preference and budget. Speaking of budget, when starting to train on flavor standards, sit down and look at what standards are likely to pop up in your brewing process and choose three to five that will be most important to you and your process. In a perfect world our panel would be able to identify 50 standards by chemical name and intensity but, while this could be your goal, this may not be a short-term reality. Keep the scope of attribute training limited and relevant to your process, which will save you money, time, and headaches.

BUDGET-FRIENDLY SUPERMARKET SPIKES

Looking for a way to dip your toes into training on flavor standards and specific attributes? There are flavors available at your supermarket or for free that represent or mimic the more precise and expensive flavor standard spikes commercially available. Experiment with varying amounts or times for each spike against a control sample until you notice a qualitative difference between the two. While supermarket flavorings may not be a perfect match for every spike, it can be a low-cost way to get your panel started training on a new skill. These spike suggestions are not intended to replace flavor standards for attribute training.

- Diacetyl: artificial butter flavor
- Isoamyl acetate: artificial banana flavor
- Acetic acid: distilled white vinegar (5%)
- Light struck: leave a hoppy beer in sun for 20 minutes
- Dimethyl sulfide: canned cream corn
- Oxidation: old product from your beer library

Up until this point you may have noticed that I have avoided the use of the term "off-flavor." This is a deliberate choice. Because modern craft brewers produce a diverse and exciting array of flavors and aromas, the term off-flavor can carry unwanted negative connotations. While a flavor attribute may be undesirable in a specific brand it may be desirable in another. Using negative terminology can distort future perceptions of different brands.

Let us look at the flavor attribute isoamyl acetate (banana ester). You may not want perceptible levels of isoamyl acetate in your stout, but if you are brewing a *hefeweizen* I am sure you want some isoamyl acetate present. So, is isoamyl acetate an off-flavor or is it simply out-of-control? This may seem like an isolated case, but isoamyl acetate, like many other standards, is present in almost every fermentation. While its presence in a stout is unwanted, I think it is more helpful to call this attribute out-of-control, which implies it can be corrected on the next go around or by some other method. This example also works for flavor attributes that are typically associated with poor yeast health, such as acetaldehyde and diacetyl. Also, knowing the source of the flavor means then you can control it.

This approach puts your sensory panel to work, because data from the panel should be used to determine what is out-of-control or not appropriate for a particular brand. The key takeaway here is only you and your panel can determine how much of a certain compound is too much for a certain brand.

Diacetyl

Diacetyl (2,3-butanedione) is part of a group of flavor compounds call vicinal diketones (VDKs), which includes 2,3-pentanedione. Diacetyl is detectable by sensory analysis at low levels, 20–60 parts per billion (ppb) in light beers, and has been one of the most talked about flavor compounds in craft beer for years. Diacetyl lends a sweet buttery and butterscotch aroma and flavor as well as a slickness on the palate that is generally an unwanted flavor attribute in most beer styles.

Diacetyl is a by-product of the yeast cells synthesizing valine, an amino acid. During glycolysis, the yeast excretes a compound into the fermenting wort called acetolactate, which is the flavorless precursor to diacetyl. This acetolactate is then chemically converted to diacetyl in the young beer via an oxidation reaction outside of the yeast cell. This reaction is non-enzymatic, meaning it occurs spontaneously and does not require the presence of yeast or its enzymes. This is an important point to remember because if you crash and remove the yeast while the acetolactate is still present this reaction will naturally proceed over time, leaving diacetyl in your beer with no way to remedy it later. If you allow the oxidation reaction to happen while the beer is still in contact with yeast, the yeast will reabsorb the diacetyl produced and convert it to products that contribute little to the final flavor of the product.

A second source of diacetyl in beer is contamination with the lactic acid bacteria *Pediococcus* and *Lactobacillus*, both of which produce diacetyl. This problem is commonly seen in "dirty" draft lines, but it can rear its ugly head just about anywhere in your brewery. Good cleaning and sanitation practices will help mitigate this source of diacetyl.

Diacetyl (VDK) Force Test

So how do you know if acetolactate precursor is present in your wort or beer? You can perform a diacetyl force test whereby you heat a sample of the beer in question, accelerating the oxidation reaction of acetolactate to diacetyl. This qualitative test is very easy to perform and can be used as a QCP. If the sample does not pass the diacetyl force test the beer does not proceed to the next step in your process.

Materials

- Erlenmeyer flask with cap or heat-resistant glass with cover
- Thermometer
- Hotplate or hot water bath

Procedure

1. Place a 400 mL sample of beer to be assessed in a 1000 mL Erlenmeyer flask. Place a thermometer into the sample and through the cap so the volatile VDK aromas do not escape.
2. Heat the sample to 150°F.
3. Remove from heat and allow to stand for 5 minutes.
4. Place the sample in an ice bath and cool to about 68°F (or close to the temperature of the beer in the tank).
5. Pour about 50 mL of the cooled sample into a glass and take another 50 mL sample of untreated beer directly from the tank.

Results

- If VDKs are present in both samples: acetolactate precursor is present; beer needs much more contact time with the yeast and should be held and retested before proceeding.
- If VDKs are present in the heated sample only: acetolactate precursor is present; beer needs more contact time with the yeast but it is close to being ready. Hold and retest before proceeding.
- If both samples are free of VDK aromas and flavors: beer is ready to proceed.

This test is most powerful when coupled with sensory training on diacetyl and a panel of two or more tasters. After introducing your panel to diacetyl at 3–6× threshold concentrations you can start lowering the spike concentration amount, for example, from 3× to 1.5× to 0.75× and so on. This will give you an idea of which panelists are sensitive tasters. They can be your go-to tasters when evaluating diacetyl force tests.

GOAL SETTING

Your sensory program is an important part of your quality control program, so much so that I encourage you to set goals and draw up a plan to execute them in a sensory analysis manual as a supplement to your quality manual. A basic example follows.

Brewery X Sensory Program

Brewery X's sensory program is intended to empower all staff to make confident and informed decisions regarding the condition and development of our beers. The program is divided into two main areas: a true-to-brand program aimed at codifying brand descriptions and statistics; and a flavor standard training program aimed at educating team members.

Program Goals

True-to-brand (true-to-type) identification is a critical skill for all members of our team, from production to tasting room to sales. Our production team requires additional expertise around fermentation assessment, water quality assessment, and go/no-go determination and production release. Finally, our program aims to provide advanced training for staff that are engaged and interested in learning more. Other general areas of focus of this program include:

- maintaining rigorous and regular training schedules,
- ensuring regular and thorough training,
- providing growth opportunities for staff,
- improving and deepening data analysis,
- identifying strengths and weaknesses in various tasters,
- empowering team members to speak up regarding sensory concerns, and
- testing and certification of sensory progress.

True-to-Brand Training

The primary goal of the true-to-brand training is to identify and codify our brand specifications, both sensory data as well as technical brewing specifications. This information is critical for production release of our beers. True-to-brand training sessions will be held twice a week on Tuesday and Thursdays at 10 a.m.

The sensory team will meet biannually to develop and review true-to-brand documentation for every brand. In addition to creating initial true-to-brand descriptions for all beers, the team will also continually monitor our existing brands over time to ensure that there is no unintentional brand drift.

The secondary goal of the sensory team is to prepare educational plans and documents necessary for true-to-brand sensory training for staff.

Documents

A current set of true-to-brand documents will be retained in the quality manual, a copy of which can be found on the production floor for reference. A second set will be retained in the tasting room for reference by Tasting Room Ambassadors. The documents will also be made available for sales and marketing staff.

SENSORY TRAINING: ENTIRE TEAM REQUIREMENTS

True-to-brand familiarity and training is the core of our basic sensory program. All team members should have good familiarity with all brands in current production.

The sensory team will strive to run at least one Entire Team Requirements training per week.

True-to-Brand Introduction

All new staff members will experience a guided tasting of current beers on draft in the tasting room as part of the new-hire onboarding training. This will be guided by a member of the Sensory Team.

True-to-Brand Identification

All team members should be able to accurately identify true-to-brand beer for all beers, with especial focus on basic flavor standards.

Basic Flavors Standards

All team members should have familiarity with and ability to identify flavor standards deemed important to Brewery X's production practices. Areas of training for all team members will include:

- acetaldehyde,
- diacetyl,
- infection—acids and acid/diacetyl combinations,
- oxidation.

Sensory analysis is a critical addition to all quality programs large and small. Not only is it an immensely valuable troubleshooting tool, sensory analysis can also be the final QCP that prevents sub-par beer from leaving your brewery. Sensory programs are relatively inexpensive to start, all you need is a beer or ingredient in a glass, a piece of paper, and an objective taster and you are good to go. After all, it is fun to drink beer, especially when you are proud of it.

14 MICROBIOLOGY

Volumes upon volumes of books have been written about brewing microbiology, for good reason. Yeast is the microbiological mechanism by which the sugar water we work so hard to make is turned into beer. Microorganisms have influenced civilizations and even created beers with terroir (think the lambics of Belgium). Only in the past hundred years or so have we been able to isolate a single species of yeast and brew consistent beer as a result. Now brewers have access to yeast suppliers who provide an endless variety of yeast strains and the tools to differentiate between them. For brewers, the tools of microbiology also put us on the offense. You no longer have to wait for a microbial issue to arise when tasting, you are proactive and make educated decisions about your beer before whatever little bugs that are living in there wreak havoc on your reputation and compromise the safety of your product. Microbiology plays a huge role in your quality system. Yes, it is hard, but I am here to tell you that you do not need to be a trained microbiologist to get good, useful results.

A basic microbiology program is an invaluable troubleshooting tool for small craft brewers. As your size and distribution grows, it becomes even more important. A simple microbiology and yeast management program was the main focus for the primary stages of building out my quality program. That started at 900 barrels—it would have started sooner if we had known how powerful the information it provided was. Several years later, we have peace of mind that the likelihood of recall or withdrawal due to a microbial issue is low. That being said, microbial issues are always a possibility and our program is still developing and improving.

WHY A MICROBIOLOGY PROGRAM?

Before we get into how to perform microbiology in your brewery, let us think about why you would want to start a microbiology program in the first place. The first reason concerns yeast, the engine that drives your brewery. You provide it with a source of fuel in the form of wort and the yeast takes care of the rest. Some brewers are expressly focused on performance, how fast they can get from OG to FG and turn that tank around to keep things moving. Others are more interested in the production of aromas and flavors the yeast contributes.

In practice, we should probably be focusing on both, striking the balance between aromas and flavors and performance. In order to do this predictably on any scale we should be concerned with the overall quality of the yeast itself. Not only does this mean the condition of our pitching yeast—cell density, viability, and vitality, all covered in chapter 12—it also means the purity of the yeast culture, that is to say, is it free of bacteria and other yeasts, domestic or wild. Unwanted microorganisms compete with the pitching yeast for nutrients in the wort. This can lead to quality and consistency issues in several ways, such as off-flavors or unwanted flavor attributes, hazes or sediments, excessively low pH, and overattenuation that can lead to gushing, overpressurization, and package failure.

Thus, the second reason for a microbiology program concerns the control of microbial contaminants, such as bacteria and domestic and wild yeast. The world as we know it is full of yeast and bacteria. As children we are taught the importance of washing our hands to prevent disease. In the same vein, our number one weapon against microbial contamination of our beer is sanitation. In addition to good sanitation practices (chap. 5), you use the microbiology program to directly and indirectly validate the efficacy of your sanitation protocols, so you can be confident that your tanks are clean and the beer coming out is uninfected.

Mold can be a particular concern for brewers. Some molds contain mycotoxins that are harmful to humans. Although I do not cover mold identification in this book, it is important to be aware of mold infections, especially when adding fruit to beer or using hard to clean/oxygen permeable holding vessels, such as when barrel aging.

PREVENTING MOLD GROWTH

- Inspect vessel/barrel for presence of mold. Clean vessel or discard barrel if present
- Limit oxygen exposure
- Inspect ingredient additions for mold. Discard if present
- If using fruit, keep below liquid level

VECTORS FOR MICROBIAL CONTAMINATION

How do unwanted bacteria and yeast end up in beer? Bacteria and yeast are all around us. They live on the skins of fruit, on brewers' grain, in the air, and on us. Keeping organisms out of your brewery is impossible but ensuring they do not take hold and overcome your pitching yeast in your fermentation vessels is not. Understanding how microorganisms get into your brewery is the first step in a successful microbiology program. If you start a microbiology program and keep finding microbial growth but you do not see any changes in your beer, you are not likely to keep up your program—there is simply too much noise. When something does go wrong, you will wish you had worked on eliminating the noise to get to the real issue.

Ingredients

Every single raw ingredient you bring into your facility has the potential to bring in contaminates.

Water. As the largest component of your beer, water can be dangerous because it can bring in microbial contaminants, particularly if there is no kill step such as boiling. Rinse water can be a possible source of contamination. Know your water sources, put it through regularly scheduled microbiological testing, and monitor for microbes that may be potential sources of contamination in your beer.

Grain. Brewers' grain is teeming with lactic acid bacteria. Think about sourdough bread, with just water and flour the dough comes to life. Likewise, in the mash tun, given some water and time, your mash will go sour (that is where the term sour mash comes from). Luckily, the mash is followed by the boil, a kill step that eliminates lactic acid bacteria. But think about milling—that dust gets everywhere, including on you. That dust can be a source of contamination or noise in your microbiology program. Set up your microbiology lab space away from the milling area; ideally, you should completely isolate the milling process from your production facility. This will lower the chances of cross contamination.

Hops. It is well known that hops have antimicrobial properties, but that does not mean hops are completely devoid of microbial life. Molds and mildews are known to live on hops in the field and they can survive into your brewery as well. If hops are stored carelessly, for example, left open to the air for periods of time or stored in contaminated secondary containers, those microbes can make their way into your brew. Nowadays hopping rates of 3, 4, even 5 lb./bbl. dry hops are not uncommon. There is no kill step in the dry hop process, so cross contamination is a huge concern when hopping at this rate. Chances are you do not pasteurize after these dry hop loads, so it is entirely possible that microbes introduced in dry hopping can wreak havoc on your packaged beer. While I have not seen this effect directly, I cannot rule it out as a possible vector of entry for microbes.

Fruit or vegetables. As stated earlier, the skins of fruit contain a plethora of yeast and bacteria. Once these microbes encounter the sugar inside the fruit or residual sugars in your beer they will go to town. Without an aggressive kill step (i.e., a vigorous boil or pasteurization) the microbes on the fruit skin will make it into your final beer. This may not be a bad thing if that is your intention, but you better be sure you can detect those microbes and control them, otherwise your beer will be packaged before it is done fermenting.

Aseptic fruit puree. A quick word on aseptic fruit purees is in order, as this ingredient has gained in popularity. While it says aseptic on the package, do you really know if it is free of microbes that can alter your product or even make the consumer sick? Did you vet the supply company and its practices? Was the puree aseptic leaving the packing facility? How was it treated on the way to your facility? How was it treated once it reached your facility? Did the person receiving the product or adding it to the fermentor drop it, or worse, did they roll it across the brewery floor? Some of these things you can see or ask. Vet your suppliers and ask for an official certificate of analysis. If a supplier cannot provide a certificate, seek one who can. Before the product arrives, write a receiving protocol and handling SOP. The larger point here is that just because it says aseptic on the package there are many factors out of your control. So, do your due diligence and determine if the product is in fact aseptic. And please, please, please never package beer that contains unfermented fruit, which is a source of sugar that will continue to ferment in package. Keeping this beer cold is not a QC program step or a disclaimer. It is up to you, the brewer/QC manager, to ensure your beer is safe for the consumer; it is not the consumer's responsibility.

Herbs and spices, process aids, or anything else added post-boil. Herbs and spices, and process aids such as kieselsol or other finings and additives—in fact, anything else added post-boil— are all possible vectors for microbes. Know how these items are treated and prepared. Before using any such addition, make sure it is listed on the FDA's Generally Recognized as Safe list of ingredients and will get formula approval from the TTB. See TTB Ruling 2015-1 and attachment-1 of that ruling for guidance (https://www.ttb.gov/beer/rulings). Just as you would with any other ingredient, write a handling protocol and ensure best practices are being followed before adding anything to your beer.

Environment

It may go without saying, but the environment in which your beer exists in and moves through can be a potential vector for microbial contamination. Even mundane everyday tasks may pose a potential risk.

Air. Yeast, bacteria, and mold live and travel in the air, making it a potential vector for contamination. Since bacteria live everywhere—in the air, on surfaces, and in water—cross contamination is a serious concern in every brewery even with closed fermentation vessels, brite tanks, and packaging lines. Brewery floors are riddled with yeast and bacteria from forklift traffic, people bringing in microbes from outside on their boots, and ingredients and waste being dumped on the floor and moved around the brewery. These things increase the potential to spread contamination. If staff are climbing into tanks they can introduce contaminants from their gloves and boots, and when cleaning up messes there is overspray from hoses, sending microbes everywhere. Cross contamination can be mitigated with good cleaning and sanitation practices.

Equipment, process piping. Equipment, particularly piping, can easily harbor microbes if not flushed and cleaned properly. Be aware of dead legs (i.e., capped pipes that have no regular flow) in your process piping. A proper flow of cleaning chemical cannot be achieved in these microbial hotspots so, over time, biofilms can form that are difficult to remove. Not every piece of equipment is fully CIP-able because there are "shadow" areas that do not get sufficient contact during CIP. Other means are necessary to clean and remove unwanted microbe from these areas.

Kegs. In today's industry, breweries are using many different strains of yeast and sharing a keg fleet. Brewers will want to make sure that their keg cleaning is effective enough to prevent cross contamination. Verify that you keg washer and keg cleaning methods are valid and working as expected. This can be done with visual inspection, microbiological screening, or ATP swabbing and testing.

> ⚠ Never attempt to open a keg without proper training and using a written SOP that has been put through a hazard assessment.

Wort aerators. Many brewers aerate their wort using compressed air. The air collected in an air compressor may be filled with unwanted microbes or machine oils. Ensure your compressed air is filtered and sanitary.

Sterilization

Sterilization is the process of making something free from living microorganisms. While boiling does kill some bacteria and fungi it will not kill all of them, so boiling is not sterilization. In the brewery lab, sterilization is easily accomplished using steam under pressure at 250°F for 15 minutes. To create this environment, you can use an autoclave or pressure cooker. I have friends who have had success using an Instant Pot® to sterilize lab equipment and media. Of course, whichever device you use to sterilize with, at least use autoclave tape, which indicates proper temperature was reached, or a biological indicator capsule to confirm sterilization was achieved.

Smaller items, such as wire inoculation loops or tweezers, can be sterilized by flame sterilization, or "flaming." Simply clean the item and then run it back and forth in a flame until it glows red. Never put the red-hot loop directly into a specimen, instead, allow the loop to cool by pressing it into media before spreading the specimen.

When sterilizing liquids, ensure the cap or closure is loose or askew to allow steam to get in and to avoid boilovers. Also be sure at the end of the cycle to relieve the pressure slowly to prevent a messy and dangerous sudden boilover due to the pressure release.

ASEPTIC TECHNIQUE

The success of any microbiology program is related to the ability of the operator to use proper aseptic technique. Aseptic technique is the use of sanitary laboratory practices to ensure sample integrity and purity during sampling, handling, and plating. This prevents contamination of cultures from foreign microbes present in the environment so a representative sample of the beer, wort, or other sample to be tested can be obtained. Without aseptic technique, you would not know if the contaminant lives on the sampling equipment, the tank, the environment, or in the sample itself.

In general, aseptic technique[1] is applied when retrieving a sample for analysis and when inoculating that sample onto culture media or some other device for biological identification. Each vessel type (fermentor, brite tank, heat exchanger, bottle, can) has its particular set of circumstances, but, once these are known, aseptic technique is not difficult to accomplish. You simply need to pay attention and be thorough. In general, you will thoroughly clean the sampling surface and sampling tools, spray with alcohol and flame until dry as appropriate, then collect the sample in a sterile vessel for analysis.

Aseptic Sampling

Aseptic sampling should be applied in your quality program whenever you are testing anything that relies on the purity and accurate representation of the sample, such as a forced fermentation or wort sterility test. The following is a general procedure to use when sampling liquids in preparation for microbiological screening.

Equipment needed

- Sterile jar, Whirl-Pak®-style bag, or other sterile receptacle
- Cotton swabs and/or brush for cleaning
- Wash bottle containing 70% alcohol
- Portable flame source, such as a lighter or butane torch
- Nitrile gloves

[1] The BA hosts training videos covering these techniques used as part of *ASBC Methods of Analysis*: "ASBC Methods of Analysis Training Videos," Brewers Association [website], accessed November 20, 2019, https://www.brewersassociation.org/educational-publications/asbc-methods-of-analysis-training-videos/.

Procedure

1. Ahead of time, prepare sampling valve or mechanism:
    a. Remove any debris (use swab or brush if necessary)
    b. Clean in alkaline solution, ensuring all points of contact are clean
    c. Rinse cleaning solution, then hold in clean sanitizing solution

If you decide to sterilize by autoclave at this stage, ensure that the sample port is sufficiently cooled so as not to kill detectable microbes during the sampling process.

2. Gather sterile receptacle.
3. Put on nitrile gloves and spray gloved hands with 70% alcohol.
4. Clean and prepare sampling surface, likely a butterfly valve.
5. Remove sampling valve from sanitizer and spray 70% alcohol inside and out of sampling valve.
6. With flame source, flame all surfaces of sampling valve that will come into contact with beer sample.
7. Flame until surface is dry.
8. Again, spray 70% alcohol inside and out of sampling valve.
9. Flame again. After a few seconds with flame still lit, let beer flow through sampling valve long enough to cool the previously heated surfaces. Collect sample in a sterile container, being careful not to touch the container opening or cap to anything that has not been aseptically treated.

When preparing your sampling valve with heat, some beer must be sacrificed to allow the port to cool down sufficiently. The initial heat from flaming the sampling valve can reduce or even eliminate the microbial population in the first beer pulled, which would result in an unrepresentative sample. Another difficulty in taking a representative sample of liquid in a tank is when samples are always taken from a single point. You can mitigate this by taking multiple samples over time, pulling samples from multiple sample points, or taking larger samples.

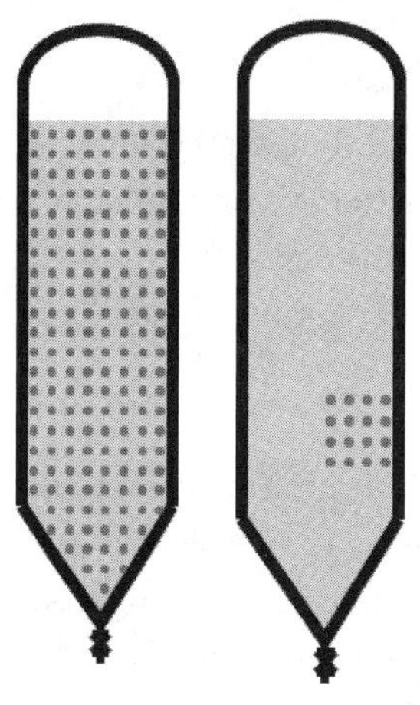

Figure 14.1. Visualization of a representative sample (on right) compared to a single sample point (on left).

Now that you have a sample that is an accurate representation of your beer, you have to apply similar aseptic techniques when inoculating the sample onto your selected culture media.

Aseptic Technique for Pouring Plates and Inoculation

Preparing a sterile work environment within your lab space is of upmost importance. Decluttering your benchtop or work area followed by cleaning and disinfecting with isopropanol are obvious first steps. Perhaps not so intuitive is creating a sterile field using an open flame. The open flame will create convection currents, also called an updraft, whereby warm air rises up and away from the flame in all directions. As the updraft rises, microorganisms present in the air are borne away, creating a sterile field in the immediate area of the flame. The flame source should be fairly small and easily controlled, like a Bunsen burner or small alcohol lamp. Applying these aseptic techniques throughout, let us walk through a general procedure to use when inoculating culture media for microbiological analysis.

You will need:

- A clean work surface
- Low-traffic area away from the busy brewery, preferably with closed doors and windows
    - A laminar flow hood is ideal and a worthwhile expense at some point.
- Wash bottle containing 70% alcohol
- Bunsen burner, alcohol lamp, or other flame source

- Sterilized
  - petri dishes
  - serological pipettes or pipette tips
  - inoculation loop or spreader
- Culture media and your samples to be tested
- Autoclave or pressure cooker

Procedure

Prepare media according to manufacturer's instructions. You will likely have to mix several grams of powdered media with several hundred milliliters of distilled water and boil for a short amount of time. Most media will have to be sterilized under steam and pressure (15 psi at 250°F).

While the media is cooling:

1. Prepare your work surface by cleaning with a detergent of your choice.
2. Wash your hands, put on nitrile gloves, and spray them down with 70% alcohol.
3. Spray your work surface with 70% alcohol and wipe down with a clean towel.
4. Create a sterile field by placing a Bunsen burner or alcohol lamp in the middle of your work surface and lighting it.
5. Gather the remaining materials needed for plating and place them nearby.

When pouring media and working under a sterile field:

1. Open media bottle and pass the lip over the flame several times.
2. Lift the cover of the petri dish just enough to allow media to enter the bottom dish.
3. Replace the cover and move the dish to a clean staging area.
4. Allow media to cool and firm up—this takes about 15–20 minutes.

When inoculating plates with an aseptic sample and working under a sterile field:

1. Open the sample container and draw up a measured amount of sample into a pipet tip.
2. Lift the cover of a cool, set petri dish just enough to allow the pipet tip to enter the petri dish.
3. Dispense the sample onto the media and withdraw the tip.
4. Sterilize your inoculation loop using the sterile field flame source.
5. Cool the loop by touching it to the media.
6. Run the loop through the sample and spread it across the media.
7. Use a fresh pipette tip for the next sample.

Take care not to completely open the petri dish to the outside air. Even under the updraft of the flame unwanted microbes could land in the dish and contaminate your results. For the same reason, do not touch the media bottle or pipette tips to the edge of the container or on the surface of the plate. These techniques are not hard to master—patience and practice will go a long way.

PLATING CULTURES

The principal means of conventional microbiological testing is the use of culture media. In the brewery microbiology lab, samples (beer, wort, yeast specimens, etc.) are used to inoculate a range of media, which is then incubated. Using selective and differential media allows identification of wild yeast and other beer spoilers.

Types of Media

General growth. General growth media supports the growth of a wide range of microorganisms. This type of media is typically used for growing a pure culture for propagation or storage.

Selective. Selective media is formulated to support the growth of one group of microorganisms and inhibit the growth of another. In the brewery lab, selective media typically contains cycloheximide for inhibiting the growth of *Saccharomyces*. Other additives promote the growth of the organisms of interest.

Differential. Differential media contains indicators that allow you to differentiate between microorganisms. A common additive to brewer's selective media is bromocresol green, a dye that some yeast strains can metabolize (the cultures grow white/colorless) and some cannot (cultures grow green). Other yeast strains have varying abilities to metabolize the dye, resulting in yellow, green, or blue cultures with varying shades in between. This can be handy for a cursory identification of yeast strains.

Selective and differential. A selective and differential medium contains an inhibitor, such as cycloheximide to select against *Saccharomyces*, and an indicator, such as bromocresol green, to differentiate microbial cultures.

In addition to choosing the type of media, the growth environment should also be considered. Some species require oxygen to produce energy and grow, while others do not. Those that require oxygen are called aerobes, those that do not need oxygen are called anaerobes (table 14.1). Manipulating the environment (e.g., reducing oxygen) in which you grow these organisms allows you to further differentiate them. Traditionally, the plated and inoculated media is placed in well-sealed jar with a gas pack sachet that produces an anaerobic environment.

TABLE 14.1 OXYGEN TOLERANCE TERMINOLOGY

Term	Explanation
Obligate aerobes	Oxygen required for growth
Facultative anaerobes	Aerobic and anaerobic growth occur but growth in presence of oxygen is greater
Obligate anaerobes	Growth stops in the presence of oxygen
Aerotolerant anaerobes	Only anaerobic growth but can survive in presence of oxygen
Microaerophiles	Require a small amount of oxygen for growth

Choosing the Right Media

There are many kinds of media available to the modern brewer. Narrowing down your choices can be difficult. You can even choose to buy many media types premade, which saves you preparation time and gives you the luxury of not dealing with concentrated chemicals. But premade media has downfalls too, such as the high cost and having to order often due to shelf life issues.

With little investment and a bit of discipline, you can prepare your own plates for less than a dollar a test. Considering the importance of a microbiology program in possibly preventing a recall, costing you a lot in terms of money and reputation, less than a dollar a test seems well worth it. The media you choose will depend on your brewing practices and the goals of your microbiology program. The more kinds of media complicated your brewing practices are the more you may want, or need, to monitor. For example, brewery X uses and re-pitches one yeast strain, so the only concern is with the purity of the culture. Let us also suppose brewery X does not have a separate milling area in the brewery. The brewers are expressly concerned with classic beer spoilers, *Lactobacillus* and *Pediococcus*, so they choose to run one media, HLP, at every step of their process.

Now consider the next example, which is from my own brewery. We run and re-pitch multiple (three to five) yeast strains year-round; we also use *Brettanomyces* in the main brewhouse. We have a barrel program and barrel storage attached to our main brewhouse, where we use multiple yeast strains and bacteria, including a coolship and a spontaneously fermented beer program. We are very concerned about the potential for all these yeast strains and bacteria to comingle. We do our best to prevent cross contamination by being mindful of the vectors mentioned earlier in this chapter. There are routine audits of cleaning processes, as well as screenings of every batch at multiple points in the brewing process. As you can see, our process has more moving parts than brewery X mentioned above, so we have added a few different kinds of media to help us cast a wider net and make sure yeast and bacteria do not end up where we do not want them.

HLP (Hsu's Lactobacillus *and* Pediococcus*) Medium*
Hsu's *Lactobacillus* and *Pediococcus* (HLP) medium is the first medium we introduced into our lab program for two major reasons. The first was the ease of identifying the common beer spoiling bacteria *Lactobacillus*

HLP: A SELECTIVE MEDIUM

- Formulated to promote the growth of lactic acid producing bacteria
- Contains cycloheximide to select against *Saccharomyces*
- Citric acid creates an acidic environment
- Sodium thioglycolate scavenges oxygen and the agar keeps a uniform anaerobic environment
- Sodium acetate inhibits the growth of Gram-negative bacteria

and *Pediococcus*, which was a major concern for us when sending beer out in kegs, bottles, and cans. Using HLP can help identify bacteria without the need for a microscope. The individual cultures that arise from a bacterium grow large enough to be seen and identified with the naked eye. Our brewery runs an HLP screening of every tank at least once a week.

The second reason is its ease of use. HLP is incredibly easy to prepare. The medium only needs to be boiled so an autoclave or pressure cooker is not required for sterilization. An anaerobic chamber or jar is not required because sodium thioglycolate scavenges any oxygen present, you can simply use test tubes that have been well cleaned and carefully sanitized. To this day, we still prepare HLP the same way as we did when we first started using it. Regardless of size, HLP should be your first choice of medium when starting a microbiology program.

Once we were able to get a routine in place running tests on HLP every week, we added a pressure cooker (autoclave) and three additional media. This greatly increased our ability to screen for other potentially dangerous and unwanted organisms.

LCSM (Lin's Cupric Sulfate Medium)

Lin's cupric sulfate medium (LCSM) is a selective medium used for detecting non-*Saccharomyces* (wild) yeast strains. The cupric sulfate in LCSM makes it a selective medium against many *S. cerevisiae* strains but not all.

Even though it is not definitively selective against all *S. cerevisiae*, LCSM can still be a powerful tool. In our brewery, we found that LCSM suppresses the growth of our house California ale yeast strain (*S. cerevisiae*), whereas "wild" yeast (*Brettanomyces*) and several other commercial yeasts grew on this media, including our *saison* and *hefeweizen* strains. This gives us the ability to detect cross contamination between these yeast strains and our California ale strain (*S. cerevisiae*) when running them side by side in our brewery.

WLD (Wallerstein Laboratory Differential) Medium)

Wallerstein laboratory differential (WLD) medium is effective at growing *Brettanomyces* and *Saccharomyces*-type wild yeast but is selective against *S. cerevisiae*. The medium is also differential because the presence of acid-producing bacteria changes the color of the

SACCHAROMYCES CEREVISIAE VAR. *DIASTATICUS*

Saccharomyces cerevisiae var. *diastaticus* (diastaticus) is a variant of *Saccharomyces cerevisiae* with the potential to spoil beer. It is feared by brewers and brew lab personnel alike, and for good reason because it is very difficult to differentiate from the *Saccharomyces* strains we rely upon to produce the beer we love.

Diastaticus is like a wolf in sheep's clothing. In addition to glucose, fructose, sucrose, maltose, and maltotriose, diastaticus has the ability to break down and ferment complex starches and dextrins that normal *S. cerevisiae* cannot. This attribute has been associated with the *STA1* gene that give diastaticus the ability to produce an enzyme called glucoamylase.* Glucoamylase is able to break off a glucose molecule from the end of a polysaccharide, allowing it to be fermented by any yeast present. This effect of diastaticus contamination can take weeks or even months to manifest depending on storage conditions and levels of diastaticus present.

As you can imagine, this is a big problem for brewers. Contamination can lead to unwanted secondary fermentation, which can result in package overpressurization due to excess CO_2. The most common reason for a market withdrawal or recall is package overpressurization, because this can be a serious danger to distributors and consumers if bottles and cans explode in storage or when handled. The secondary fermentation can also push your alcohol levels out of TTB compliance, and lead to flavor changes and gushing beer.

There has been a lot of recent coverage on potential diastaticus contamination of yeast obtained from suppliers, but it is entirely possible for diastaticus to be introduced to your facility through vectors such as the ones discussed in this chapter. In their study covering 2008 to 2017, Meier-Dörnberg et al. were able to trace 71% of diastaticus contamination events to the bottling line. Screening for diastaticus using traditional plating methods is challenging but well worth your time and effort.

* Tim Meier-Dörnberg, Fritz Jacob, Maximilian Michel, and Mathias Hutzler, "Incidence of Saccharomyces cerevisiae var. diastaticus in the Beverage Industry: Cases of Contamination, 2008–2017," *MBAA TQ* 54, no. 4 (2017): 140–48.

TABLE 14.2 MEDIA SELECTION BREAK DOWN

Media name:	What it detects:	Why to choose this:
HLP (Hsu's *Lactobacillus/Pediococcus* Medium)	*Lactobacillus* and Pediococcus	Ease of use and differentiation
LCSM (Lin's Cupric Sulfate Medium)	Non-*Saccharomyces*-type wild yeast	Some *Saccharomyces* types will grow
LMDA without cycloheximide (Lees Multi-Differential Agar)	General media	Acid-producing bacteria change media color; yeast strains produce markedly different morphologies
WLD agar (Wallerstein Laboratory Differential Agar)	Bacteria, wild yeast, mold	Selective and differential

medium, which can be seen without even turning on the microscope. Thus, WLD means we can suppress the growth of our house ale yeast strain to detect other yeast species and allows us to easily see a change in color on the media that indicates we need to get out our Gram staining kit (see p. 137 below); all other colonies go under the microscope for further investigation.

LMDA (Lee's Multi-Differential Agar)
Lee's multi-differential agar (LMDA) without cycloheximide has the potential to pick up most brewery related yeast and bacteria. While there are other general types of media available, such as SDA, UBA,[2] and wort agar, I prefer LMDA because colony morphologies are markedly different between most of the different yeast strains we use regularly in-house. This gives us the ability to recognize our house strain morphology versus some other kind of growth. While not a perfect method for identifying many kinds of yeast or bacteria, LMDA can be a useful tip-off to investigate further and double-check other media we used to screen the same batch.

Building Confidence in Your Results

Positive and negative controls. Positive and negative controls ensure the validity of the test. This involves running control tests for positive and negative results alongside your experimental test. All media prepared in your microbiology lab should have a negative and positive control.

A **negative control** group is a plate or tube that is not exposed to the experimental treatment (i.e., not inoculated with sample). The negative control should have no growth whatsoever, which tells you that aseptic technique during preparation of the media was performed correctly and that the plates or tubes were in fact sterile when filled.

A **positive control** is not exposed to the experimental treatment but is exposed to a known contaminant that is known to produce the expected effect. The source of the contaminant will depend on the media used. For example, a great positive control for HLP media is just a few specks of grain dust from your mill. Floors and drains are typically successful positive controls for all other bacteria and wild yeast media. You could also be more deliberate and use a stock of pure cultures appropriate for positive controls with various media.

Force testing. When introducing a new medium you will want to validate that it performs as you expect. We call this force testing. In order to do this, you will want to obtain a known sample of what you are trying to force. For example, when commissioning LCSM we obtained *Brettanomyces bruxellensis*, a commercially available non-*Saccharomyces* strain, to see if it would grow. Once *B. bruxellensis* growth on LCSM was confirmed, we trialed dosing lower and lower cell densities to see how low the threshold of detection was for our lab. In other words, the lowest number cells of *B. bruxellensis* per milliliter in a plated sample that showed growth on the media. In another example, when choosing a general growth media, we plated our house ale yeast and all the yeasts we typically bring in and then compared the results. We saw that the colony morphology differed between our house ale yeast and some of the other

[2] Schwarz Differential Agar and Universal Beer Agar.

yeasts we bring in regularly. We then cataloged the different yeast strains in a photo database for help identifying them in future tests.

Force testing is also useful when bringing in new yeast to see how the pure culture behaves across all the media you use for your routine microbiological screening. Since yeast pitches are so dense you will want to perform a sterile serial dilution. Sterilize the diluting water in the autoclave (or pressure cooker) and use aseptic technique in as many steps as necessary to reduce the cell density to 300,000 cells in 1 mL of sample. Then simply document what the growth (or lack thereof) looks like for each medium.

Redundant and repeat testing. You may notice that there is some overlap between medias and what they detect. This redundancy is purposeful and helpful for validating results. For example, if you see growth in your HLP tubes, you will likely see corresponding results on anaerobically incubated WLD. If you do not see corroborating results you will likely want to repeat testing to see if you can find consistency in growth between media and maybe put a QC hold the beer. Repeat testing (i.e., testing in duplicate or triplicate) also builds confidence in your results. Lack of consistency in repeat samples may suggest an issue with technique. Multiple inoculations using different sample sizes also helps validate your testing method; for example, if you inoculate samples of 1 mL, 2 mL, and 3 mL into HLP you should essentially see 1×, 2×, and 3× colonies formed, respectively

Plating Methods

Spread plate method. The spread plate method is a suitable technique with any kind of agar media. A small liquid sample (up to 0.2 mL) is pipetted onto an agar plate and a sterile cell spreader is used to distribute the sample evenly around the entire plate. Proper spreading of the sample should result in visible and isolated colonies of yeast or bacteria that are evenly distributed on the plate. A plate count can quantify the number of total microorganism cells present in the sample. This number is expressed in **colony forming units** (CFUs), which is based on the assumption that, theoretically, each visible colony could have arisen from a single viable cell.

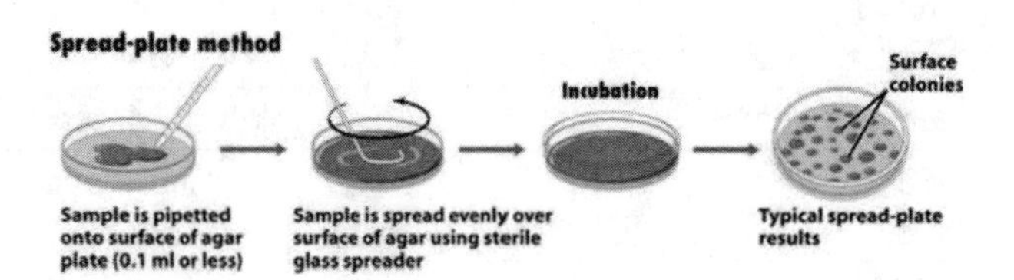

Figure 14.2. Illustration of the spread plate method.

viable cell count (CFU/ml) = (no. of colonies × dilution factor) / volume inoculated on culture plate

If the sample was not diluted, then the dilution factor equals 1.

As an example, suppose you performed a 100× serial dilution and plated 0.2 mL of sample onto WLD agar. After incubation you counted 20 colonies counted.

viable cell count = (20 colonies × 100) / 0.2 mL
= 10,000 CFU/mL

From this you estimate there are 10,000 viable cells per milliliter. You can look closely at the colonies and their morphologies; if they are markedly different your sample may be a mixed culture. At this point you would make up more media and transplant each unique colony and use streak plating to isolate the species.

Streak plating. Streak plating results in individual microbial cells being spread far enough apart that it results in pure culture colonies forming.

1. Use a flame-sterilized (cooled!) inoculation loop to carefully pick an unique isolated colony from the previous spread plate culture, then spread it over the first quadrant of a fresh plate (approximately one-quarter of the plate) using close parallel streaks moving from the edge of the plate toward the center.
    a. Alternatively, you can start with a small liquid sample (up to 0.2 mL), which is deposited near one edge. Then spread this sample the same way as before with a sterile loop.
2. Re-sterilize the loop in the flame and cool it by touching it to the agar at an unused edge of the plate. Pass the loop through the already streaked first quadrant once. Using the same close parallel streaks, spread the sample over the second quadrant.

3. Re-sterilize and repeat step 2, one or two more times as room on the plate allows. Cover and incubate the plate.

Successive streaking of the sample across different quadrants of the plate dilutes the sample to the point where there is only one cell deposited every few millimeters, so the resulting colonies will be pure cultures.[3] Once a pure colony is isolated, you can take further steps in identifying the microorganism by propagating it or sending it out to a third-party lab.

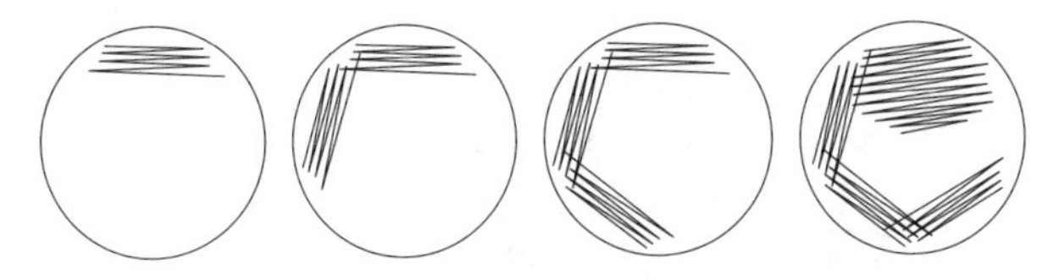

Figure 14.3. Illustration of the streak plate method.

Pour plate method. The pour plate method differs from the spread or streak plate methods because the sample being tested (inoculum) is added to a petri dish before the warm agar medium is poured on top. As with any plating method, there are advantages and disadvantages to pour plating. An advantage is that this method allows you to use a larger sample (1–10 mL), which gives a better snapshot of what is going on microbiologically in the tank. Also, the mixing of the sample and agar media can allow growth of more oxygen-sensitive microorganisms (e.g., *Lactobacillus acetotolerans*). The disadvantage is that organism growth will be suppressed if media is poured on too hot. Close attention must be paid to the temperature of the media, with the optimal temperature being 100–113°F. This temperature window can close fast, so inoculating a large sample set is difficult without a hot water bath. When performing your plate count on a poured plate be aware colonies can grow throughout the media (not just at the surface) and colony size may be reduced, so take care not to overlook the small colonies.

[3] See BA's training video "ASBC Methods of Analysis Training Videos," https://www.brewersassociation.org/educational-publications/asbc-methods-of-analysis-training-videos/.

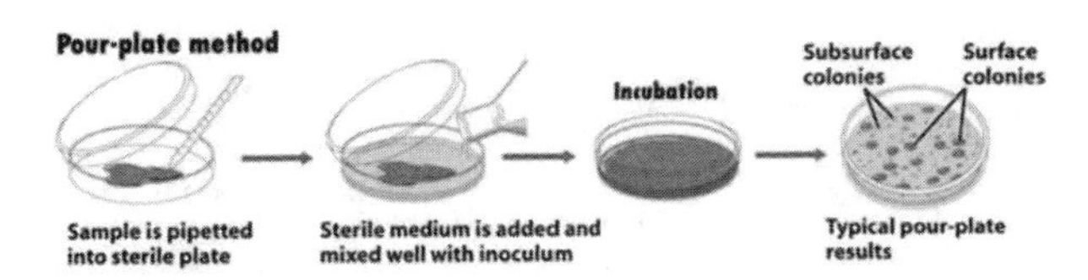

Figure 14.4. Illustration of the pour plate method.

Membrane filter plate. The membrane filter plate method requires some additional equipment, a filter funnel and vacuum system. These setups can vary in price, but a simple setup can be found for around $150. Membrane filters have a uniform porosity of predetermined size (generally 0.45 μm) sufficiently small to trap microorganisms. Using the vacuum filter setup, 10–100 mL of sample is drawn through the membrane filter. This provides a larger sample size, which is better when screening for small populations of spoilage organisms and thus lends itself well for screening water sources and packaged product. After an adequate sample is passed through it, the membrane filter is transferred onto media in a 60 mm petri dish and incubated.

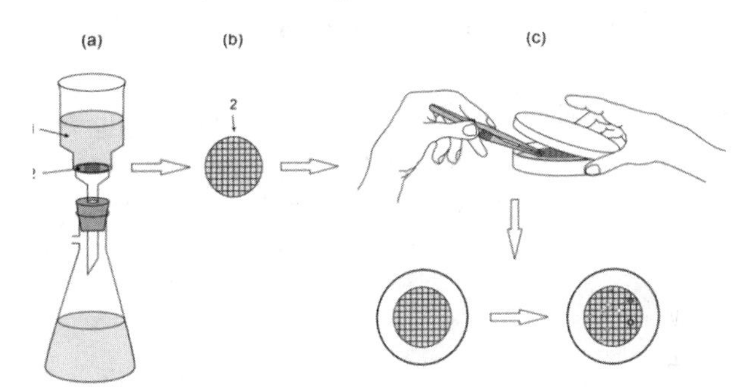

Figure 14.5. Illustration of the membrane filter plate method.

Enrichment broths. Some populations of microorganisms can be present in your beer but below the threshold of detection. This can be a scary realization when you are sending beer far and wide or if someone decides to cellar your beer. Enrichment broths can turn a minority population into a majority population by providing specific nutrients to the organisms of interest. They are especially useful when paired with additional selective components. For example, De Man, Rogosa and Sharpe (MRS) broth with cycloheximide is effective at growing up lactic acid bacteria (*Lactobacillus* and *Pediococcus*) while suppressing yeast. After a microorganism has been enriched in broth media, plating methods can be employed for enumeration or identification.

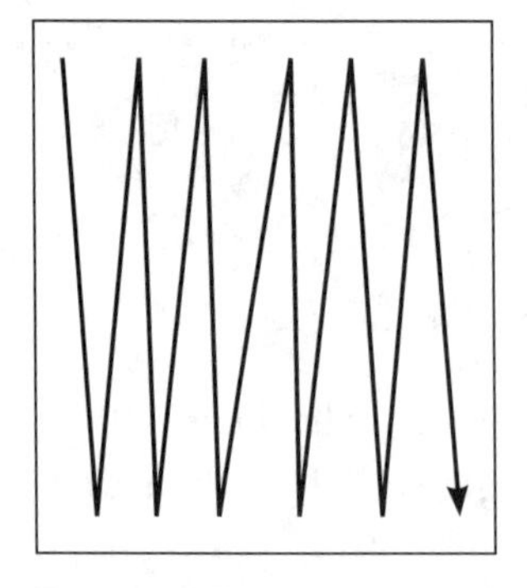
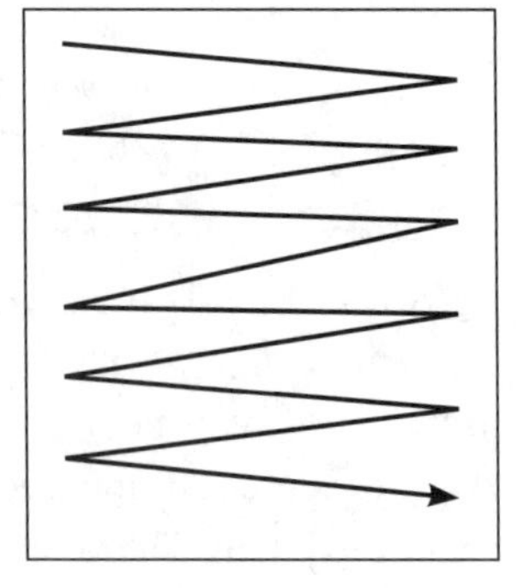

Figure 14.6. The swabbing pattern used for the sterile swabbing method.

Sterile swabbing. Surfaces can be tested directly by using a sterile cotton swab to test a surface. Sterile swabs should pre-moistened in the lab with sterile water, then placed in a sterile tube or Whirl-Pak and brought to the location of interest. Choose a 10 cm × 10 cm surface to swab. In some cases this may not be possible, but do your best to get as close to that surface area as possible. Use parallel strokes, from right to left (horizontally) and then from top to bottom (vertically), across the sampling area to collect a representative sample, rolling the swab between the fingers while you do this. Aseptically place the swab back in the sterile container and bring to the lab to use it to inoculate your media. Sterile swabbing is a useful method to directly test a surface in question, such as hoses, equipment parts, or the inside of tanks.

Check plate testing. Check plate testing, also called air testing, is used to determine the microbial activity present in a particular area of your brewery or some other environment. Cooled and set media is exposed for a predetermined amount of time in the area of interest; it is then covered and incubated. The resulting growth will give you an idea of what microbial activity you are encountering in that area. Check plate testing can be a useful tool when troubleshooting issues that arise. You can check the efficacy of your sterile technique and the sterile field you are creating with your flame by setting check plates at different distances as you work. Another useful experiment with check plates is to place them around your packaging line when not in use, leaving them there over the length of a normal canning run.

I HAVE GROWTH ON MY PLATES!

Unexpected growth on your media can be a scary thing, but it does not necessarily mean the batch needs to be dumped. You will want to do everything in your power to collect as much information on the microorganism and ultimately do your best to identify the growth. Just like we established with our quality planning in chapter 7, you will want to come up with an SOP and assign responsibility for carrying out these tasks. In our lab, the order of operations is as follows:

1. Initiate a hold on the affected product according to the in-house product hold and release SOP.
    a. Retest the sample in question
2. Pluck a colony from the original plate and re-culture by applying to a streak plate.
3. Mount and observe cells under microscope, using 40× magnification for yeast and 100× for bacteria.
    a. If yeast, note the cell size, shape, and budding characteristics (skip to step 7, and reference the Yeast Identification section below).
    b. If bacteria are found, proceed to step 4.
4. Gram stain and record results
5. Catalase test and record results
6. Oxidase test and record results
7. Compile the results and hit the books to identify the microorganism.
    a. If results are concerning, initiate a QC hold.
    b. If the results from the retest are repeated:
        i. Send out samples to a third-party lab for confirmation of results.
        ii. Decide on the fate of the beer in question.

Yeast Identification

Yeast strain identification is generally hard to do without sophisticated equipment such as PCR, which is possible in your lab and I will briefly discuss this later in the chapter. In our lab, we send samples out to a third-party lab whenever there is an issue with yeast growth. While awaiting the results, I will try to learn more about the problem using what information I can attain myself using the lab microscope, such as budding patterns, along with other clues gleaned from the type of media the yeast grew on, and re-culturing the yeast to determine its fermentation requirements.

Saccharomyces growth. Differentiating *Saccharomyces* species will likely be the most difficult because they may look like your house strain in many ways. It is a good idea to create a photographic record of your house yeast strain's morphology on streak plates. This record is also a good tool for training new lab personnel. When trying to identify an unknown *Saccharomyces* yeast, I look for differences in colony morphology and color compared to documented growth patterns I have compiled while working with several yeast strains, both known and unknown, on general media. I will look for budding patterns under the microscope and compare them to known patterns of that yeast. While I may not be able to specifically identify the yeast in question, I can at least tell that it is not what I intended to pitch.

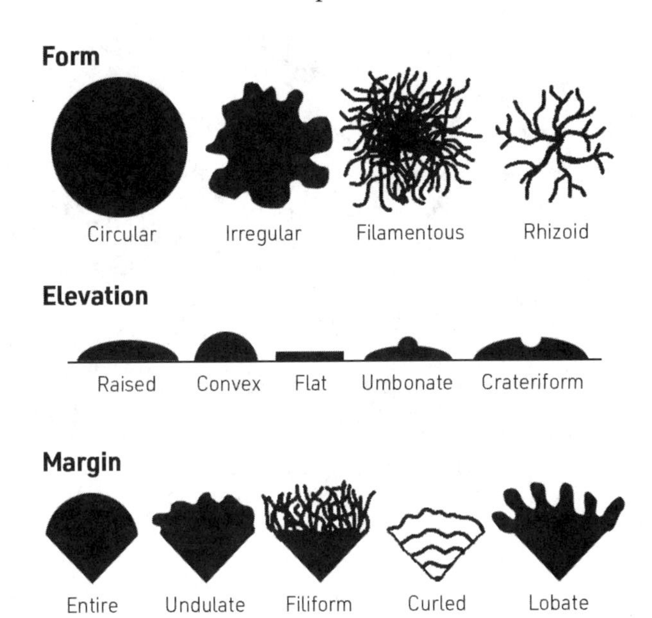

Figure 14.7. Depiction of colony morphologies and terminology used to describe them.

Non-Saccharomyces growth. With suspected non-*Saccharomyces* growth, I will again pay attention to which media it grew on and colony and microscopic morphology. An important question, is it cycloheximide resistant? Wild yeast is likely to be, while *Saccharomyces* strains typically are not. The budding pattern can suggest a certain genus of wild yeast but is not necessarily a definitive marker.

Ultimately, the genus of yeast you detect does not really matter, what matters is whether it will continue to ferment your beer or push it out of your true-to-brand specifications. You can get an idea of how the isolated yeast will behave by growing it up in wort and running a trial fermentation or forced fermentation on a benchtop scale. Documenting these traits can help cross reference the PCR analysis results you obtain either in house or from a third-party lab.

Bacteria Identification

Bacteria identification is usually less challenging than for yeast. In our brewery, after re-culturing and isolating a bacterial colony, we will look to growth on plates and in tubes for clues as to its identity. Using traditional methods based on morphology, staining methods, and enzymatic reactions (metabolism) help with identification. Three of the most common investigations are Gram staining, catalase testing, and oxidase testing.

Gram staining. Developed by Danish scientist Hans Christian Gram in 1884, the Gram staining procedure is a tool used to classify bacteria based on their forms, cellular morphologies, and reactions to Gram stain. The procedure consists of a series of stains and reagents that react with bacterial cell walls, the results of which can be analyzed under the microscope. Gram staining is not useful for identifying yeast because yeast universally stain the same when treated with Gram stain, telling us nothing.

The Gram staining procedure uses a primary stain, crystal violet, which is taken up into the cell walls of the bacteria. The crystal violet is trapped in the cell walls by the addition of iodine, which forms a crystal violet–iodine complex. A decolorizing agent is then applied and it is at this stage that the difference between the cell walls of bacteria becomes apparent. Some species of bacteria have a thick outer layer of peptidoglycan (protein-sugar complex) that makes up around 60%–90% of their cell wall, while other species only have a thin peptidoglycan layer sandwiched between two other layers. When exposed to decolorizer, those bacteria with the thick peptidoglycan layer become dehydrated, which closes the pores in the cell wall and causes the crystal violet–iodine stain to remain trapped in the thick layer of peptidoglycan. Thus, these bacteria remain stained a purple-blue

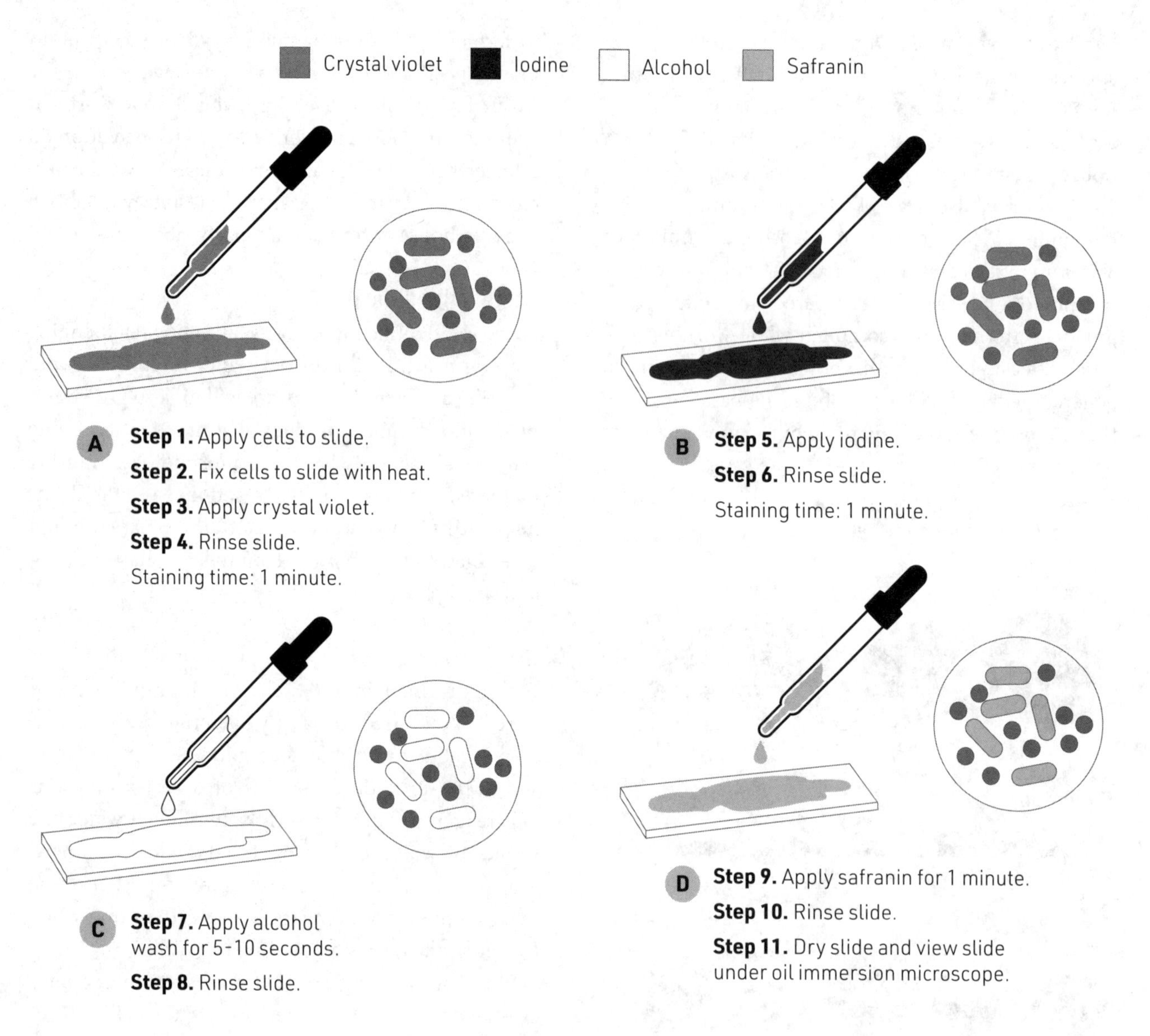

Figure 14.8. The steps of the Gram staining procedure.

color and are said to be "Gram-positive." By contrast, those bacteria with the thin peptidoglycan layer have their outer cell wall stripped away by the decolorizer, washing away most or all of the crystal violet–iodine stain and exposing the thin peptidoglycan layer. These bacteria are said to be "Gram-negative."

After decolorizing, a counterstain is applied, usually the compound safranin. Safranin stains all bacterial cells, allowing the Gram-negative bacteria to be easily seen under a microscope as pinkish-red in color. The Gram-positive bacteria are still deeply stained with crystal violet, so the pink color does not show up on them and they remain purple-blue in appearance under the microscope.[4] Classic beer spoilers, such as *Lactobacillus* and *Pediococcus* species, are Gram-positive bacteria. Wort spoilers are typically Gram-negative bacteria, the presence of which can indicate water or sanitation issues.[5]

To perform a Gram stain, a thin layer of bacterial growth is taken from your media and applied to a glass slide for investigation under the microscope. This can be accomplished by wet mounting the bacteria in distilled water and allowing the sample to air

[4] "Gram Stain," Microbugz (website), accessed June 6, 2019, https://www.austincc.edu/microbugz/gram_stain.php.

[5] Mary Pellettieri, Quality Management: Essential Planning for Breweries (Boulder: Brewers Publications, 2015), 73..

dry on the slide. The bacteria should then be fixed to the slide with gentle heat, but not too much otherwise this will cause morphological changes.

When your sample is under the microscope, you will observe three major characteristics: cell morphology, arrangement pattern of the cells, and the cell color after Gram staining. There are two major morphologies used to identify bacteria. The first are rod-shaped bacteria called bacilli, as in *Lactobacillus*. The second are spherical (i.e., generally spherical, ovoid, or round in shape) bacteria called coccus, as in *Pediococcus*.

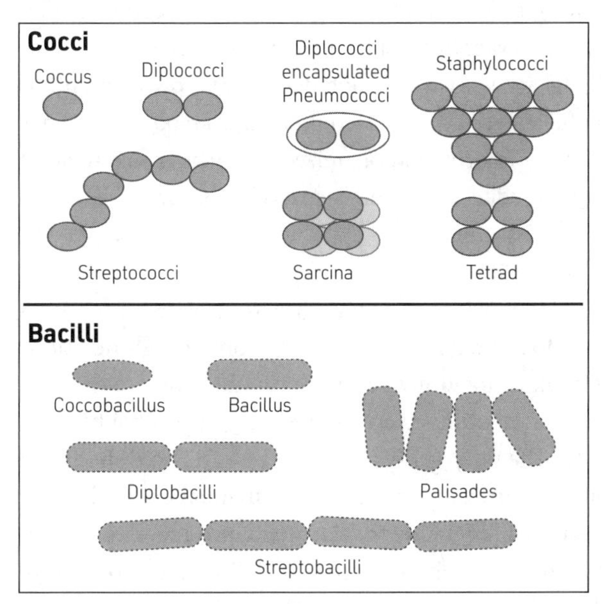

Figure 14.9. Depiction of bacteria budding patterns and terminology used to describe them.

Catalase test. The catalase test demonstrates the presence of the enzyme catalase in your sample. Catalase breaks down hydrogen peroxide (H_2O_2) into water (H_2O) and oxygen (O_2). This enzyme is employed by nearly all organisms that live in oxygenated environments as a defense mechanism against common reactive oxygen species. Many of the major beer spoiling microorganisms are "catalase negative," making the catalase test a perfect complement to confirm the results of Gram staining.

To find out if a particular bacterial colony has catalase activity, pluck a small colony and smear it onto a glass slide or in a test tube, then apply a drop of 3% hydrogen peroxide solution (which can be found at any pharmacy). If you observe the rapid formation of bubbles (oxygen) the sample is positive for catalase. The absence of catalase is evidenced by lack of or only weak production of bubbles. Catalase-positive bacteria include strict aerobes and facultative anaerobes that can respire using oxygen. Catalase-negative bacteria may be strict anaerobes, or they may be facultative anaerobes that can tolerate oxygen but only ferment and do not respire using oxygen. This piece of the puzzle helps us differentiate between bacteria that are morphologically similar but use oxygen differently.

Oxidase test. Another rapid enzyme test, the oxidase test, determines whether the bacteria in question produces the enzyme cytochrome c oxidase. The test is useful for characterizing Gram-negative bacteria. Generally, oxidase test materials come in the form of liquid phenylenediamine indicator reagent or a test strip/swab impregnated with the reagent.

To perform a catalase test, pluck an isolated colony and rub it on the reagent-inoculated medium (test strip or reagent-soaked filter paper). When the bacteria come in contact with the indicator reagent, any cytochrome c oxidase will oxidize it and turn it purple in color in only 10–30 seconds. This is an oxidase-positive result, which means the bacteria possess cytochrome c oxidase and can therefore use oxygen for energy production. No color change within 60 seconds indicates an oxidase-negative result, which means the bacteria do not use cytochrome c oxidase and do not use oxygen for energy production.

Carbohydrate fermentation test. The carbohydrate fermentation test is used to determine whether bacteria can ferment a specific carbohydrate source. Carbohydrate fermentation patterns are useful in differentiating among bacterial groups or species. Test tubes filled with a single carbohydrate source (glucose, sucrose, or maltose), a pH indicator (phenol red), and an inverted Durham tube are inoculated with an isolated bacterial colony. After incubation, the results are interpreted as follows:

- a color change, caused by a drop in pH, is a positive result for acid production;
- no change in indicator color is a negative result for acid production;
- a bubble in the Durham tube is a positive result for gas production;

- no bubble in the Durham tube indicates a negative result for gas production.

After performing all this testing, you have gathered quite a bit of information about the suspect microorganism. You can take this information to an identification chart to give you an idea of what kind of beer or wort spoiler it is (fig. 14.10).

Document Your Results

If you are running any test on a beer at any stage it should be recorded. When performing microbiological testing on each batch of beer, a microbiology report should be generated. *Even if the results are negative they should be recorded somewhere.* This documentation will aid you in verifying sanitation protocols and will be tremendously helpful when troubleshooting. The way you document your reports can vary greatly depending on the kind of testing you are doing. It can be as simple as a field item included in your brew log. This should include the date tested, negative (<1 CFU) or positive (>1 CFU) results, and notes on colony morphology.

If you are using more than two media or testing at several locations, you may want to consider a separate form for microbiological reporting (fig. 14.11). Each report should follow each brew from beginning to end. This should include:

- Important tracking information
    - Batch, gyle number, and brew date
    - Brand
    - Fermentation vessel
    - Yeast pitch information (type, harvest date and source, generation number)
- Wort sterility test results
- Info for all locations tested
    - Where the sample was taken
    - Who took/plated the sample
    - Date it was plated
    - The media it was plated on and method
- Results from the media, including incubation conditions (e.g., aerobic or anaerobic)

To supplement your microbiological documentation, you may want to consider an identification matrix. In this digital age it is easy to photodocument growth and meaningful test results and insert them into a report using links. Keeping information organized in a matrix like this can help expedite results and keep track of recurring issues.

Next Steps

Now that you have your growth isolated, identified, and documented, there may be some tough decisions to make. Hopefully you have caught it before the beer hits distribution. If so, the next step is probably a QC hold. The nitty-gritty and scope of a QC hold should be outlined in your written quality manual but at the very least should include the date held, product description, quantity held, and reason held. Additionally, the entire staff should have clear instructions of how to treat/proceed with the product in hold.

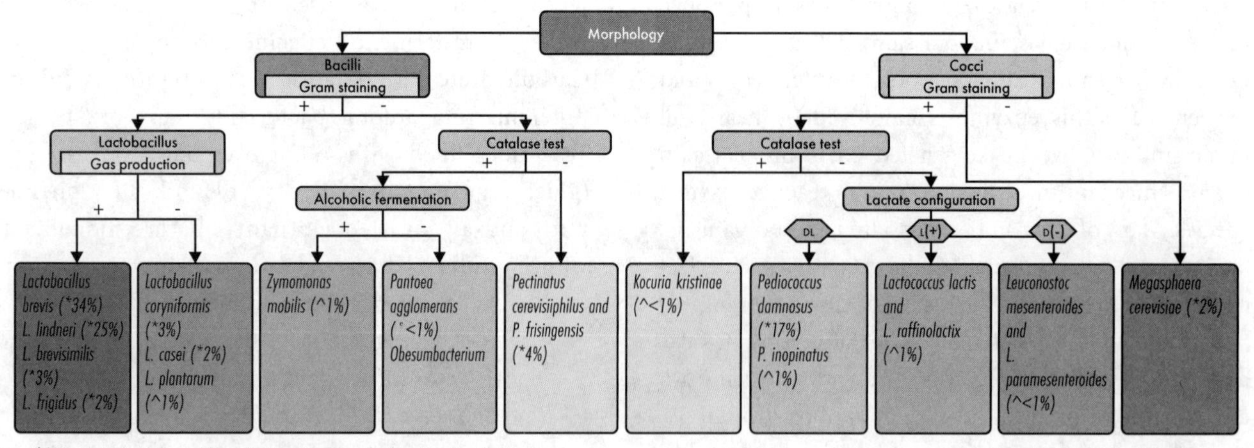

Figure 14.10. Identification flow chart for beer spoilage bacteria. © Blüml, Susanne, and Sven Fischer. 2004. *Handbuch der Fülltechnik: Grundlagen und Praxis für das Abfüllen flüssiger Produkte*. Hamburg: Behr.

Batch ID:	Brew Date:	Fermentor:
Brand:		

Wort Sterility	Gyle #	24 hr	48hr	72hr	Notes:
	Plate if positive result (sign of ferment or turbid)				If negative, discard after 72hr

Micro screening	Control point	FV	Date tested		Technician
	Media	48 hr	96 hr	Aerobic or anaerobic	Notes:
	HLP				
	WLD				
	LCSM				
	LMDA				
	If positive result, perform Gram stain, catalase, and oxidase test				

Micro screening	Control point	BBT	Date tested		Technician
	Media	48 hr	96 hr	Aerobic or anaerobic	Notes:
	HLP				
	WLD				
	LCSM				
	LMDA				
	If positive result, perform Gram stain, catalase, and oxidase test				

Micro screening	Control point	Package	Date tested		Technician
	Media	48 hr	96 hr	Aerobic or anaerobic	Notes:
	HLP				
	WLD				
	LCSM				
	LMDA				
	If positive result, perform Gram stain, catalase, and oxidase test (if applicable)				

Identification	Test	Result	Notes:
	Gram stain		
	Catalase		
	Oxidase		

Figure 14.11. Making your microbiological reports as traceable as possible will help build confidence in your results as well as assisting with root cause analysis when problems occur.

The QC hold will allow time for the QA/QC department to investigate fully. Gather as much information as possible to make an informed decision regarding the fate of the beer in hold. Further testing should include, but not be limited to:

- Retesting the beer in question, trying to repeat the results
    - Interpreting retest results:
        - Same growth or different (increasing/decreasing CFU)?
        - Were additional growth conditions included (aerobic, anaerobic, etc.) and what were the results?
- Force aging the beer in question (incubate and see what happens)
    - Monitor QC parameters, especially sensory, gravity, CO_2, and pH
- Sending out a sample for third-party testing
- Comparing findings with microbiology records using past documentation and ID matrix
- Performing a root cause analysis. Try and determine why it has happened and institute policy changes to prevent it from reoccurring.

After examining the results of all your testing, you should be armed with enough information to make an educated decision regarding the fate of your beer. There are several approaches you can take depending on the spoilage potential of the microorganism present. If there is potential for overattenuation, I advise waiting for the beer to fully attenuate before packaging. Overattenuation can lead to a multitude of problems, including package overpressurization and gushing, as well as driving your beer out of true-to-brand ABV specification. If the beer in question is already packaged, then it should be discussed as to whether the product will become dangerous to the consumer and should be destroyed instead. Another factor to consider is if the organism's presence will put the beer out of product specification before the end of its shelf life. This could be bad for your reputation if the consumer has expectations of a product and it does not deliver in one way or another. Ultimately, I cannot tell you what the best decision is for your brand. Look to your quality priorities written in your quality plan to aid you with this difficult decision. Regardless of your decision, make sure you do your due diligence in figuring out what the problem was and work toward swiftly correcting it.

POLYMERASE CHAIN REACTION (PCR)

The polymerase chain reaction (PCR) and associated techniques allow for rapid detection and identification of microorganisms, though it does require a trained technician and is fairly pricey. Not many small breweries will be able to include PCR in their lab set-up, but it is a valuable resource to aspire to as your brewery grows. PCR amplifies the DNA of your suspect microorganism using short nucleic acid sequences that are unique to the DNA of the microorganism in question. It is highly sensitive—you can use PCR to detect if even one copy of the target DNA is present. Since each organism has its own unique genome, PCR is an incredibly powerful tool for the identification of potential beer spoiling yeast and bacteria.

PCR takes advantage of DNA polymerase, an enzyme that replicates new DNA using an existing DNA strand, which is known as the template. This replication process occurs in all living cells but in PCR we are forcing it to happen in a test tube. In order for PCR to produce useful results we have to be selective about which region of DNA is to be copied. To do this, we use short pieces of single-stranded DNA, called *primers*, that identify specific regions of DNA. Two primers are used in each PCR reaction, and they are designed so that they flank the target region, that is, the region to be replicated. Each primer is designed to have a sequence that will make it bind to the template DNA, just at the edge of the region to be copied.[6]

Now, imagine our test tube. We have our ingredients to make copies of the DNA region of interest. First, we have a strand of template DNA that we are trying to copy, which is from our organism of interest. Second, we have a DNA polymerase; in our case the heat-stable polymerase *Taq*, originally derived from the heat-tolerant bacteria *Thermus aquaticus*. Third, we have our primers known to be unique to the DNA sequence of the organism of interest. Fourth, we have free nucleotides—the building blocks of DNA—for the polymerase to assemble DNA copies from. Finally, a buffer solution is used to provide the correct environment. We can now "cook up" our ingredients to force the specific

6 "Polymerase Chain Reaction," Kahn Academy (website), accessed August 20, 2019, https://www.khanacademy.org/science/biology/biotech-dna-technology/dna-sequencing-pcr-electrophoresis/a/polymerase-chain-reaction-pcr.

reactions we are looking for and facilitate successive rounds of replication of our template DNA. This is done with a thermocycler.

A thermocycler is a device that has a thermal block where tubes holding the PCR reaction mixtures can be inserted. The thermocycler then raises and lowers the temperature of the block in discrete, preprogrammed steps. There are three main steps that are repeated 20–40 times (fig. 14.12).

1. **Denaturing:** the double-stranded template DNA is heated to 94–95°C, which causes the DNA to separate into two single strands.
2. **Annealing:** the temperature is lowered to between 50°C and 56°C, which enables the primers to attach to the single strands of template DNA.
3. **Extending:** the temperature is raised to 72°C and the region of template DNA between the two primers is replicated by *Taq* polymerase.

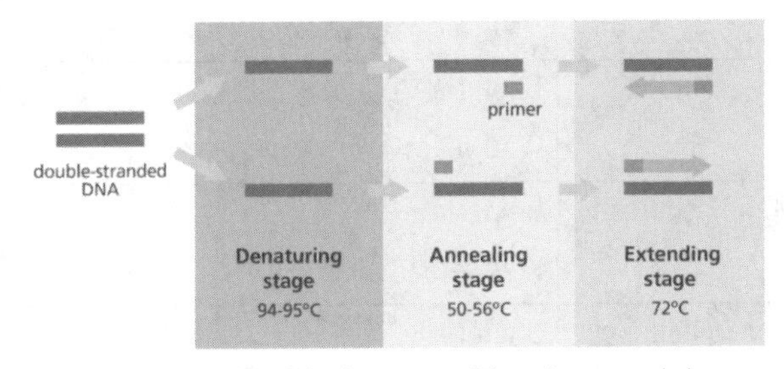

Figure 14.12. Illustration of the three steps of the polymerase chain reaction performed on a DNA sample in a thermocycler.

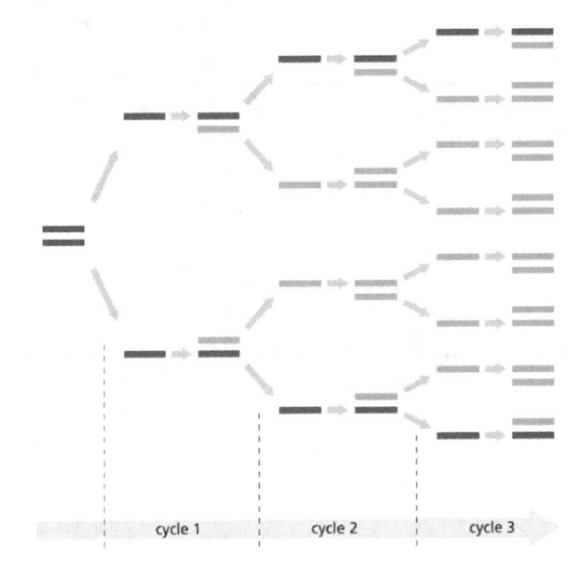

Figure 14.13. Replication of DNA over three full rounds in a thermocycler with a successfully matched primer.

After each cycle, the DNA copies serve as new templates for *Taq* to work on, so the number of templates doubles each time. This results in the amount of the specific DNA region increasing exponentially, so it becomes amplified enormously in a relatively short amount of time (fig. 14.13).[7] Using the PCR reaction, we can grow up a single strand of DNA into millions of strands rather quickly, whereas traditional plating could take days and days with non-specific results.

Bringing PCR into Your Brewery

PCR can support a brewery microbiology program by providing fast, reliable identification of beer spoilage microorganisms. A great place to start is with commercially available PCR kits. The kits come with premade and validated primers, preprogrammed thermocyclers, and step-by-step instructions to help ensure success. Common PCR kit targets are *Lactobacillus*, *Pediococcus*, wild yeast (*Brettanomyces*), *Saccharomyces cerevisiae* var. *diastaticus*, *Megasphaera*, and *Pectinatus*. Make sure you are getting the coverage you need out of your kit, especially if it is a closed platform and you are stuck with that kit's proprietary equipment, reagents, primers, and programs.

While the principles of PCR do not vary substantially from kit to kit, the methods by which you analyze your data can. Many popular kits utilize end-point PCR, where the template DNA is amplified in the thermocycler and placed on agarose gel or some other means for visualization, such as a test strip. The appearance of indicator on the test strip will imply a positive or negative result as well as the relative quantity of the template DNA.

The second and more versatile method of analysis is by quantitative PCR (qPCR). You may have heard this referred to as real-time PCR, because the PCR products are detected continuously throughout cycling. Thermocyclers that are used for qPCR are equipped with fluorescent light detectors. You add an intercalating (SYBR Green I) or probe-based fluorescent dye to your PCR reaction mixture. As the target DNA is amplified so is the fluorescent signal. Using a standard curve generated by a known amount of target DNA, the amplified DNA can be quantified. When shopping for a qPCR system, consider leaning toward an open platform, which will allow you to use kits from many suppliers and will give

7 "What is PCR (polymerase chain reaction)?" Yourgenome.org, accessed July 25, 2019, https://www.yourgenome.org/facts/what-is-pcr-polymerase-chain-reaction.

you the option to move to making your own assays and primers as you get more comfortable with the technique. This could save you a tremendous amount of money in consumables down the line.

While an immensely powerful tool for the brewing microbiologist, PCR should not replace traditional microbiological culturing but should be used to support it. Although it takes longer, culturing on various media can cast a wide net and is useful to validate and verify CIP systems. PCR is very specific and can identify only the DNA you are targeting, making it well-suited to confirming growth from culture media after traditional identification methods have narrowed your search. Practicing these microbiology methods will help you be confident that the beer you send out is the beer you intended to present to your consumers. See appendix H for a list of equipment to get your microbiology program started.

SAMPLING PLAN

TABLE 14.3 MICROBIOLOGY CONTROL POINTS

QCP	When and how often	What you are looking for
Water: incoming municipal water line and post-treatment water line	Monthly	Bacteria Wild yeast
Yeast propagation: check for pure culture **Commercial pitch:** check for purity of culture	Before pitching	Bacteria Wild yeast
Wort cooling: wort sterility test, plate growth	Each brew	Bacteria Wild yeast
Fermentor	Each brew, preferably after 24 hours of residency time	Bacteria Wild yeast
Brite beer tanks	Each brew, preferably after 24 hours of residency time	Bacteria Wild yeast
Package	Every run	Bacteria Wild yeast
Post-CIP	Each CIP program; tested once a month	Residual cleaning chemical ATP Bacteria Wild yeast
Environmental: identify problem areas flagged by check plates	Monthly	Bacteria Wild yeast
Library beer	Monthly	Bacteria Wild yeast

15
PACKAGING

In the brewery, you put a tremendous amount of care and energy into brewing and cellaring your beer in order to make it the best you can. But for distributed beer the physical package you choose and its condition is equally as important as keeping DO low and nailing your CO_2 volumes. The keg, can, or bottle is the final resting place for your product before it makes its way out of your hands and into the wild. And it can be a wild marketplace—not every person outside of the brewery will treat your beer with the respect you do. Making sure your product leaves in the best condition possible will help ensure your consumers get the best representation of your beer.

SELECTING AND RECEIVING PACKING MATERIALS

Just like brewing, packaging starts with selecting raw materials. Careful consideration should be paid to selecting the appropriate packaging material. Every brewer will have their own preference as far as materials go, but no matter what format is chosen the brewer should understand the pressure limitations of the vessel. Before receiving your packing materials you should request a technical drawing of the bottle/can/keg. This should include the precise specifications needed to seal the vessel as well as the pressure limitation that should not be exceeded. Keep in mind that as your beer warms up the pressure of your vessel will increase, pushing the vessel to its limit.

Once an appropriate vessel is chosen, there should be a procedure in place for receiving the bottles, cans, or kegs you will be filling. Examine the physical condition your packaging arrives in. Are there dented cans or broken/defective bottles? Are the slip sheets in good condition, free of soil or mold? Often forgotten is to assess the delivery truck for foul or off odors. Vapor transfer is a real thing and can impart

BOTTLE DEFECTS

Bottle defects vary in nature and severity. They often are the packaging defect most detrimental to consumer health and safety. Breaks, foreign object inclusion, chips and cracks, rough surfaces, hairline fractures, birdswings, stones, blisters, and other glass defects can result in personal injury or be a potential hazard to consumers. Get to know what these defects look like and work toward finding and eliminating them so that they do not make it into consumers' hands. Developing procedures surrounding broken glass on and around your bottling line and removing defective bottles before they are filled are important first steps in developing food safety or HACCP plans.

unwanted flavor attributes to your final product. Have an SOP in place that addresses the overall condition of packaging materials and raw materials alike when receiving and what to do when something out of the ordinary happens. When building your food safety program, assess for control points surrounding the use of your packaging materials.

FILL HEIGHT

Fill heights are important for several reasons, the first one being DO pickup. The lower the fill height the larger the headspace. With a large headspace the potential for oxidation is greater, thus, your product degrades unnecessarily quickly. Just as important is customer perception. With a severely underfilled package the consumer will feel shortchanged and that will leave a bad taste in their mouth (likely from oxidation) and them less likely to buy your beer again. Conversely, an overfilled package is harder and messier to open, which is also a bad look and suggests to the consumer that you do not really know what you are doing. Under or overfilling can also land you in trouble with the TTB regarding taxation if there are discrepancies in the amount of beer you are actually selling.

If you are running a high-speed packaging line of some sort, you will likely want to invest in a full bottle inspection system that can detect fill height, cap position, and mislabeled product as well as reject out of specification containers. The likelihood is that you are running a packaging line that is a reasonable speed for operators to do some of these functions manually. For manual inspection, determining fill height simply requires the package material and an accurate scale.

1. Determine the tare weight of the package by weighing (in grams) at least 10 empty cans with ends or bottles with caps
2. Determine the weight of desired fill in grams of water.
3. Determine the specific gravity (SG) of the beer to be packaged.
4. Multiply the weight in grams of water fill by the beer SG to convert to grams of beer.
5. Add the tare weight of the package to the result of step 4 to get the target fill weight.

Let us go through a quick example:

- Average tare weigh of 16 oz. can is 15 g
- Fill volume is 16 fl. oz.
- A fluid ounce of water is equal to 29.6 mL, which weighs 29.6 g
- Final beer is 1.009 SG

$$16 \text{ fl. oz.} \times \frac{29.6 \text{ g}}{1 \text{ fl. oz.}} = 473.6 \text{ g} \qquad \text{(calc. for step 2)}$$

From this, we can determine the weight of beer in a filled can:

$$473.6 \text{ g} \times 1.009 = 477.9 \text{ g} \qquad \text{(step 4)}$$

$$477.9 \text{ g} + 15 \text{ g} = 492.9 \text{ g} \qquad \text{(step 5)}$$

Note that, due to the manufacturing process, bottles may vary in weight. This can make it difficult to get an accurate tare weight for bottles and you may find yourself having to recalculate your package tare weight every time the lot changes. Can weights do not vary as much as bottles, so cans may have to be rechecked less frequently.

CLOSURES

Once filled, we must close that package up. A flawlessly sealed bottle or can ensures safety and assures better quality. Leaking cans or bottles are indicative of poor seaming or capping operations. A leaking container is a bad look to your distributors, retailers, and consumers. Not only is that one leaking container unsaleable but it can render surrounding beer unsaleable as well. Leaking beer will damage the outer packaging material used, rendering it soggy and useless. It will rust cans and crowns and weaken the integrity of other containers around it. The leaked beer can cause mold and invite other unwanted organisms to the party, leaving rancid and other unwanted odors on the packaging, which will eventually color consumers' evaluations of your beer. You can mitigate this damage by routinely measuring your closures and performing preventive maintenance on your closing systems.

Can Seam Measurements

Proper can seam evaluation is crucial to the successful operation of any canning machine. With the proper attention and correct training, staying on top of your seam integrity is not complicated. Double seam specifications should be obtained from your can supplier and should be adhered to in order to prevent leaks.

How to Measure Can Seams

Manual seam evaluation. Manual seam evaluation requires knowing the specifications set by your can manufacturer and use of the following tools:

- a can seam micrometer,
- a Vernier caliper (also called a caliper micrometer), and
- a pair of side cutters (or modified can opener).

Even if you opt for a more high-tech method for measuring, it is important to know how to manually evaluate seams accurately in case your equipment goes down.

Video seam imager. Available from several manufacturers, video seam imagers typically come as a package. A package includes hardware for physically measuring and manipulating the can to be imaged and measured digitally, and the accompanying software. Video seam imagers can digitally capture countersink, seam thickness, and optically obtain cross-sectional measurements of the body hook, cover hooks, seam height, and overlap. The supplied software will allow you to easily create reports for analysis and record keeping.

X-ray seam scope. An X-ray seam scope is a nondestructive option that performs both cross-sectional measurements (the same as a video seam imager provides) and a 360° tightness scan. All of this is captured digitally for your analysis using software provided.

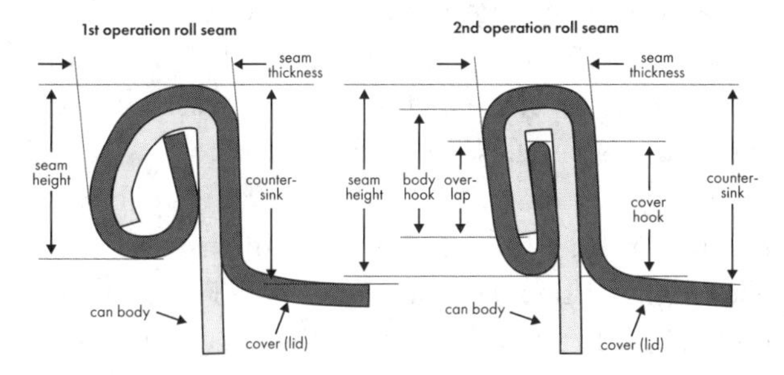

Figure 15.1 Cross-sectional view of double seam operations.

First operation external measurements

- Countersink depth
- Seam thickness

First operation internal evaluation (if available)

- The formed end of the can body and the formed end of the cover are parallel to each other and the can body wall
- The body hook is formed into the primary seal area and the overlap is well established

Second operation external measurements

- Countersink depth
- Seam thickness
- Seam height

Second operation seam tear down measurements

- Body hook
- Cover hook

Tightness rating

Tightness rating is a critical seam evaluation. The tightness rating refers to a visual evaluation of how well looseness wrinkles are removed or "ironed out" from a seam.[1] Loose seams in cans are a common cause of leaks. In order to evaluate the tightness rating the cover hook needs to be removed from the rest of the seam and lid. This is often referred to as a seam "tear down." A tear down in preparation for a tightness rating evaluation can be completed manually with end cutting nippers, wire cutters, or can opener. It is important not to distort the ends when performing the tear down—practice makes perfect. Once the cover hook is removed the tightness evaluation can be performed. The rating is expressed in increments of 10% (fig. 15.2). If no wrinkles are present, the seam is rated 100% tight; if the wrinkle extends 40% of the way down the cover hook, the seam is said to be 60% tight, and so on. The minimum tightness rating for a typical aluminum can is said to be 90% tight.[2]

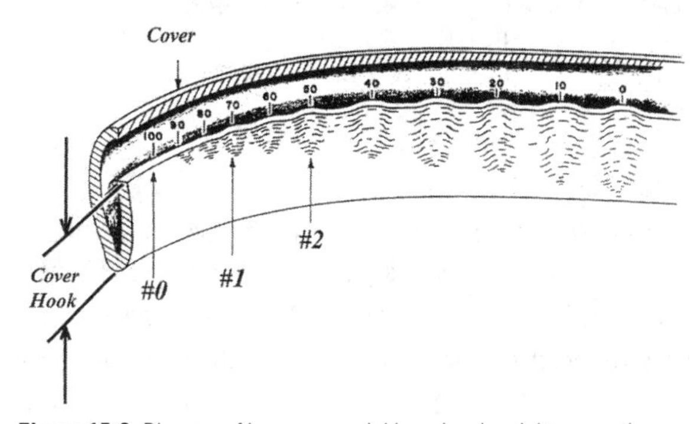

Figure 15.2. Diagram of looseness wrinkles, showing tightness rating increments of 10%. Image courtesy of OneVision Corporation.

---

1 Ray Klimovitz and Karl Ockert, *Beer Packaging*, 2nd ed. (St. Paul, MN: Master Brewers Association of the Americas, 2014), 109.

2 Klimovitz and Ockert, 109.

Can Seam Sampling Plan

There are many factors that will influence the frequency of container seal inspection, such as general company policy and the type of crimper or seaming mechanism. As far as can seams are concerned, you should conduct a complete evaluation and tear down of each seaming station prior to operation, after the machine warms up, after a major jam, and for at least every four hours of operation. It is a good idea to monitor external second operation measurements (i.e., countersink depth, seam thickness, and seam height) at timed intervals throughout the run. If seams are found to be drifting out of specification a full tear down should be executed. If a deviation is found, it is a good practice to confirm the out-of-specification result with another seam before making adjustments to the seamer. Keep a record of all your measurements because they will be of use to the next operator and can help inform them of the need for preventive maintenance.

CAN LINERS

Brewers should be aware that not all beer types are compatible with all types of beer cans. There are a multitude of can liners available and not all are suitable for every kind of beer. A mismatched beer and can liner combination can lead to unwanted sensory outcomes or liner performance issues. Brewers who package beer with low pH (<4.3) or high alcohol (>7% ABV) should make sure their liners are compatible. The Brewers Association has a resource available to members at https://www.brewersassociation.org called "Ensuring Quality and Compatibility of Beer Packaged in Cans" for those interested in learning more.

Bottle Seals

A crown crimp gauge, sometimes called a go/no-go gauge, is a simple tool used to measure the crimping of the crown after a bottle is filled and sealed. The gauge consists of a series of precise holes drilled into a thick metal sheet. Typically, the two center holes (marked "GO") represent the lower and higher allowable tolerance; the two outer holes (marked "NO-GO") represent crimped crowns that are either too small or too big for the bottle. These gauges are available online and are relatively inexpensive.

1.120 crimp gauge

1.125 crimp gauge

1.130 crimp gauge

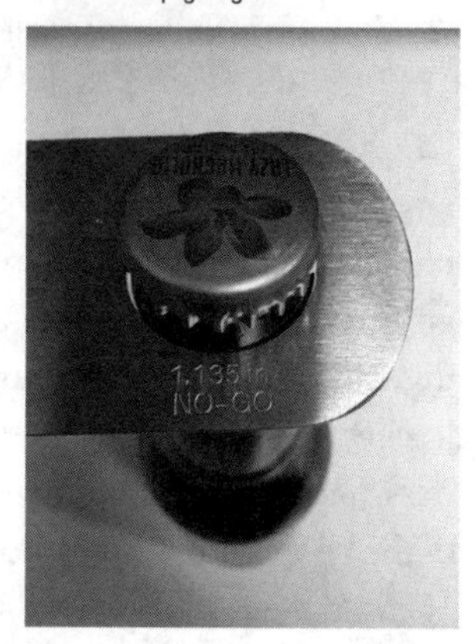

1.135 crimp gauge

Figure 15.3 Using a crimp gauge to assess bottle crowns.

When purchasing a crown crimp gauge, check with your bottle manufacturer for the correct sizing. For example, for a bottle with a 26 mm opening the appropriate crimp gauge will have holes sized 1.120 in., 1.125 in., 1.130 in., and 1.135 in. In this example, if you can pass the crown of a sealed bottle through the smallest hole the seal is too tight and the integrity is compromised. Your crown should fit through at least the 1.125 or 1.130-inch holes in the gauge with little resistance for a "go." Therefore, the crown should fit easily through the largest (1.135 in.) hole; if it does not, the crown is too loose and will not hold a seal (fig. 15.3).

If packing off manually, visual assessment of the crown should be done at pack-off. Check to see if the crown is seated straight and if there is any unusual wear around the crown that could indicate rubbing and may indicate the need for preventive maintenance.

Secondary Tests

In conjunction with seam testing, there are a few things you can do to test the integrity of your packaging.

Hot water bath. Placing cans or bottles in a hot water bath raises the internal pressure of the package and forces leakage. If you find leakers, it is time to take a closer look at the failure point.

Drop test. A drop test is for cans only. Take a can from the hot water bath, bring it outside and drop it on its end. You will expect to see the seamed end deform but not fail.

Pressure check. You should retain some bottles or cans after packaging day and check for a pressure drop. A drop in package pressure will indicate there is a gas leak in the crown or seam.

DATE CODING

It is in the best interest of brewers, distributors, importers, and consumers that all beer from small and independent craft brewers be identifiable by some form of a date or lot code on the packaged beer.[3] It is the responsibility of all brewers to ensure that their products are clearly coded for traceability. If a seam fails, a package becomes overpressurized, or there is a recall of one of your raw ingredients or process aids, the code is the mechanism used to get that potentially dangerous or poor-quality beer off the shelves.

Ultimately the method of coding is up to your brewery resources. Methods range from handwritten numbers to laser printed dates and codes. I believe transparency is best for the brewer, consumer, distributor, and retailer, so a simple, clear born-on date is best for cans, kegs, and bottles. When physically applying dates, know your market and where your beer ends up. If you are selling cases out of the brewery directly to consumers maybe a simple grocery store pricing gun and sticker will suffice. But be wary, if you ever do need to recall beer the regulatory authorities have the right to remove every drop of beer that is inadequately coded. This could mean every drop of your beer without a code will be pulled from the market.

The complexity and spread of your distribution chain should be matched by the robustness of your date coding application method. In an industry where growth can happen fast, do not outgrow your date coding and traceability methods. Growth can come to a halt if old product gets into the hands of the consumer or if you are unable to retrieve a large batch of bad beer.

CONTAINER LOT TRACKING

As part of your lot tracking and traceability efforts, it is important to keep track of the lot number of the packaging material that the beer went into and on what date. Packaging material may have an undetected defect that can adversely affect the quality or safety of your product down the road. There may even be a recall for one reason or another on that packing material. Being able to locate that packaged beer down the road is important.

The example below is a simple but effective way to track which batches make it into which lot of cans or bottles. Associate the package lot with your unique batch tracking number and the date it was filled. Once the pallet is finished, keep these records in a centralized file to call up in the future if necessary.

Can Lot Tracking		
Can lot code	**Date**	**Batch number**
67J-1000-9567	2/12/19	156
	2/13/19	157
	2/14/19	158
	2/19/19	160
48C-5566-7878	2/21/19	159
	2/22/19	161

LABELING

Labeling should not be overlooked. Ensure that your label is correctly affixed to your container. Set a list of criteria for operators to check and ensure compliance.

1. The beer in package should be correctly identified by the package label.
2. The label should be on straight, with no bumps or deformities.
3. Check that the printing registration is acceptable and the label is clearly printed otherwise.

3 "Industry Updates, Date Lot Coding," Brewers Association (blog), February 1, 2019, https://www.brewersassociation.org/brewing-industry-updates/date-lot-coding/.

You can add any criteria relevant to your packaging process as a final check before sending the beer out. The sharper your package looks the better the impression of the quality inside the package.

Beer packaging is a quality parameter some take for granted, and it does not end with a cool-looking label. Sometimes the package is the only interaction that people will have with your brewery. Making sure that the package is filled to the correct height, the beer is sealed up tight, and there is a date code will go a long way with consumers' and retailers' perceptions.

TABLE 15.1 PACKAGING CONTROL POINTS

QCPs	When and how often	What you are looking for
Receiving	Upon arrival	Visual inspection of condition; detecting foul or pungent chemical odors
Handling	Each transport step	General condition
Labels	Throughout run	Label matches the product; application appearance; absence of
Date coding	Throughout run	Code matches product; code is legible
Crowns and seams	Test before run or immediately after starting up; timed intervals throughout the run	Closures within manufacturer's specifications
Secondary tests: Drop test Hot water bath	After startup or after adjustments to seamer	Seam failure; leaking product
After packing day	Scheduled intervals	Check pressure for CO_2 change

16

CARBON DIOXIDE

Carbon dioxide (CO_2) is a clear, colorless gas and is a small, but key, component of our atmosphere. Carbon dioxide dissolves very easily in water and is found dissolved in oceans, lakes, glaciers, and all other bodies of water, liquid or frozen. It is also a by-product of your metabolism and you exhale large quantities of it every day. Just like in nature, carbon dioxide plays many roles the brewery.

We refer to the presence of CO_2 in beer as carbonation, those tiny little bubbles that fill our beer with life. In fact, after water and ethanol, CO_2 is the most abundant chemical constituent in beer. Carbonation plays a major role in our sensory perception of beer, most notably in the mouthfeel. A correctly carbonated stout, for example, can be smooth and luxurious, whereas an overcarbonated stout can be astringent and biting. That is not to say highly carbonated beers are bad. A highly carbonated *saison* is a beautiful thing, the carbonation boosting the aromatics from the yeast and drying the beer on your pallet.

While for quality purposes we are mostly interested in the CO_2 dissolved in beer, around the brewery we also take advantage of other characteristics of CO_2. For instance, we use CO_2 counterpressure to push beer from one tank to another to avoid foaming; or, because CO_2 is heavier than air, we can use it to displace other gases (like oxygen) out of tanks, cans, bottles, and other vessels. As useful and common as CO_2 is in the brewery, I would be remiss if I did not mention how dangerous this wonder molecule can be—see the hazards sidebar.

At the brewery, we are typically interested in the carbonation level of our beverage. Hitting our targets for brand and style are crucial, not least because it means our accounts can pour our beer properly. Either way you look at it, understanding how to get CO_2 into your beverage will help build confidence when you come to measure it.

CARBONATION LEVEL

Henry's Law

When beer is in a closed vessel, such as a brite tank, can, bottle, or keg, the concentration of CO_2 dissolved in the beer will reach an equilibrium. When at equilibrium, the partial pressure of a gas (in our case CO_2) equalizes between the liquid phase (in our case, beer) and the gas phase (in the headspace). When at equilibrium, we can calculate the concentration of CO_2 in liquid using Henry's law. Henry's law states that, at a constant temperature, the concentration of a dissolved gas in a liquid is directly proportional to the pressure of that gas in the gaseous phase.

$$P_{gas} = H_{gas} \cdot M_{gas} \qquad \text{(Henry's law)}$$

Where:

P = absolute gas pressure

H = Henry's law constant for gas under equilibrium conditions

M = molar fraction of the gas in beer

Note that if the beer is dispensed under pressure, say, from a keg, then absolute gas pressure is atmospheric pressure (14.7 psi, i.e., 1 atm) plus the gauge pressure.

At low temperatures a gas is more readily soluble in solution. At high temperatures a gas is less soluble and, therefore, requires a higher absolute pressure to remain in solution at a given concentration. This is why Henry's law states *at a constant temperature*. The Henry's law "constant" (H) is not a true physical constant, it must be determined experimentally for a gas at a given temperature. Within the temperature range typical for dispensing beer, H increases as temperature increases. This can be seen when a keg is not chilled to serving temperature and pours foam. The pressure required to push the warm beer and keep CO_2 in solution is not sufficient and the foaming you see is the effect of CO_2 breaking out of solution.

The volume of headspace is also a consideration when carbonating, as this affects the partial pressure of the gas in solution. The partial pressure refers to what portion of a mixture is a specific gas. Thus, if the partial pressure of CO_2 contained within the gas of the headspace is greater than the partial pressure of gas dissolved in the beer, the CO_2 will continue to dissolve into solution until equilibrium is reached. If the partial pressure of CO_2 is greater in the beer, the gas will break out of solution in order to equilibrate in the headspace lowering the carbonation of the beer.[1]

Time is another factor to consider when carbonating a beer. Henry's law applies to a system that has reached

[1] Steven R. Holle, *A Handbook of Basic Brewing Calculations* (St. Paul, MN: Master Brewers Association of the Americas, 2003), 65.

CARBON DIOXIDE HAZARDS

Carbon dioxide is colorless and odorless. Because CO_2 is heavier than air, it displaces oxygen when it builds up in enclosed areas, which can lead to a serious asphyxiation risk. Although CO_2 causes a burning, acrid sensation in the nasal passages, this symptom should not be used as a means of detecting the gas. Even short-term exposure can lead to dizziness, loss of consciousness, and death. See permissble exposure limits in the table below. Stay safe—invest in an indoor air quality (IAQ) meter and install it in your cellar and any other confined area, such as a walk-in cooler used for beer dispense.

Occupational Safety and Health Administration (OSHA): Guideline CO_2 Permissible Exposure Limits for General Industry	
8-hour time-weighted average exposure limit	5000 ppm (0.5% CO_2 in air)
Short-term exposure limit	30,000 ppm (3.0% CO_2 in air)
Immediately dangerous to human life	40,000 ppm (4.0% CO_2 in air)

Source: OSHA Occupational Chemical Database, s.v. "Carbon Dioxide," Occupational Safety and Health Administration (website), accessed October 2, 2019, https://www.osha.gov/chemicaldata/chemResult.html?recNo=183.

At standard pressure and temperature CO_2 is a gas and behaves as such, meaning it has no definite volume or shape but will simply fill whatever volume is available to it. Adding CO_2 gas to whatever tank or container you are pressurizing will increase the pressure exerted in all directions onto any liquid present and the walls of the vessel. This can lead to the pressure exceeding the limitations of your container, resulting in failure by rupture explosion, which can lead to serious injury or death. Pressurizing closed containers is serious business. Ensure the container is rated for the pressures you are using and do not allow fermentation to continue inside a closed vessel with no release valve.

Carbon dioxide gas will also react with common chemicals used for CIP procedures, such a caustic (NaOH). This results in an endothermic reaction that can cause a tank to implode. Properly venting CO_2 will reduce the risk of implosion and increase the efficacy of the chemical used in CIP.

equilibrium. If you are using constant head pressure alone to carbonate your beer it can take a long time for the partial pressure between the gas and liquid phases to equilibrate. The limiting factor here is the interface between the surface of the liquid and headspace, in other words, the surface area to volume ratio. A molecule of gaseous CO_2 can only dissolve in the liquid when it comes into contact with the liquid.

To speed up the carbonation process brewers use a carbonation stone made of sintered stainless steel or porous ceramic. The stone is inserted into the beer in the tank and CO_2 forced through it. This produces fine bubbles of CO_2 on the outside of the stone, increasing the surface area of beer that interfaces with CO_2. As long as the partial pressure of CO_2 in the headspace is greater than the partial pressure of CO_2 in the beer, the CO_2 supplied through the stone will readily (and rapidly) dissolve into the beer.

Units

Carbonation is measured as either "volumes" or grams CO_2 per liter. One volume is equivalent to one liter of uncompressed CO_2 dissolved in one liter of beer, which is equivalent to 1.96 g CO_2/L. Beer carbonation can range from 1.2 volumes in cask beer all the way up to 5.0 volumes in a bottled wheat beer (2.4–9.8 g/L). Most beers fall between 2.0 and 3.0 volumes (3.9–5.9 g/L).

$$\frac{1.96 \text{ g } CO_2}{1} = 1 \text{ volume } CO_2$$

WHERE TO MEASURE

Calculation of CO_2 as Related to Extract

Carbon dioxide is a product of fermentation, specifically, the anaerobic respiration of yeast. However, when we look at the equation for the fermentation of sugar to ethanol (C_2H_5OH) and CO_2, you can see that only 46.4% of the fermentable sugars are turned to CO_2.

$$\begin{array}{ccccccccc} C_6H_{12}O_6 & = & 2C_2H_5OH & + & 2CO_2 & + & \text{yeast growth} & + & \text{by-products} \\ 100\text{ g} & = & 48.6\text{ g} & + & 46.4\text{ g} & + & \approx 3\text{ g} & + & \approx 2\text{ g} \end{array}$$

The reason 100% of the mass balance does not go to ethanol and carbon dioxide is because the sugar is used for other metabolic and growth pathways. In any case, the above equation tells us that, depending on the fermentable extract in your wort, 1 bbl. of wort can yield roughly 4.5 kg (10 lb.) of CO_2. The exact amount of CO_2 produced can be calculated using the OE in degrees Plato and the RDF. Craft brewers are not typically concerned with measuring CO_2 at this point though, because recovery systems are cost prohibitive at that relatively small scale.

As you may have gathered by now, we are expressly concerned about measuring CO_2 dissolved in solution. This essentially means the beer in the brite tank and packaged beer. Two major methods of forced carbonation are employed in the craft brewery. The first method is where CO_2 is injected into into the headspace of a tank until equilibrium is reached at a constant temperature and pressure; this can take some time. Mentioned earlier, the second method is where CO_2 is injected through a sintered stone either in-line on the way to the tank or in the brite tank itself. CO_2 is supplied until the desired carbonation level is achieved. This method of carbonation is industry standard and can be relatively fast, accurate, and repeatable.

You can also elect to naturally carbonate beer through trapping CO_2 produced during fermentation, achieved by capping a tank before the end of fermentation or by priming, including priming by krausening. Both methods require precise density measurements of the remaining fermentable extract and a thorough understanding of the safety risks involved.

CO_2 Volumes as a function of residual extract
1°P (1.004 SG) ≈ 2.5 vol CO_2
0.1°P (1.004 SG) ≈ 0.25 vol CO_2

Brite Beer Tanks

Quite possibly the most important QCP is when you determine CO_2 volumes before beer gets put into package. You should not use brite tank pressure and temperature readings alone to determine CO_2 volumes. Any instrument capable of at-line measurements should suffice. Be aware of unwanted break out of CO_2 to get an accurate reading. Break out is an indication that there is not enough pressure in the system to keep the CO_2 in solution. Include a carbonation parameter in your go/no-go sensory assessment. While you probably cannot taste the difference between 2.5 volumes and 2.56 volumes, I am sure you can tell the difference between 1.5 and 2.5 volumes and you will know that the beer is not going to package flat.

When bottle conditioning, an often-overlooked measurement is the CO_2 reading before dosing with priming sugar, a significant factor when calculating the amount of priming sugar needed to reach the target CO_2 volumes. Ensure that the instrument you choose is capable of measuring CO_2 volumes at such low ranges. If krausening or capping at the end of fermentation, taking a reading of the existing carbonation level will be helpful in order determine how much krausening beer to add or when to close your vessel (be sure it can handle the pressure generated).

Kegs

Determining the CO_2 volumes in kegs is an important and useful QCP. First, it confirms your brite tank measurements as well as validates you successfully filled the vessel. Second, and particularly important, is that CO_2 content is the determining factor in balancing a draft system. While craft beer bars have come a long way, the majority of restaurants and bars still have systems balanced to pour beers around 2.5 volumes and very limited capabilities for balancing. Undercarbonated beers will pour flat at first and eventually will become carbonated to the system balance over time. An overcarbonated keg will pour foamy and be difficult or even impossible to pour if physical corrections to the system are not made to get it back in balance. Most bars do not possess the capability to adjust the balance of their draft system, so it is up to the craft brewer to ensure kegs are carbonated to the correct volume. This will make pouring easy with little waste and increase the chances an account will reorder your beer time and time again. Before freaking out about reports of foamy kegs, be sure to check the bar's serving temperature because this is a major factor in system balance (most are balanced for 38°F) and the commonest cause of foaming. For more information about draft systems and the factors surrounding them check out http://www.draughtquality.org.

Bottles and Cans

One of the final quality checks before beer hits the market is measuring CO_2 volumes in package. Depending on your packaging process, you may notice package CO_2 volumes do not completely agree with your brite tank volumes. Do not fret, your instruments are working fine—there is typically a loss of CO_2 in packaging. Agitation and a raise in temperature, which are inherent in most packaging processes, usually account for this loss. Most breweries will compensate for this loss by slightly increasing brite tank CO_2 volumes before packing. Recording the final CO_2 value in package and checking it intermittently (from your library beer) can tell you a lot about the condition of your beer over its shelf life, such as the condition of the container seal and if there was refermentation after packaging day.

CO_2 Purity

A 99.90% CO_2 purity grading for beverage-grade gases is now mandated by the Food and Drug Administration. A qualified gas provider should be able to provide you with certification of your gas's purity. If you want to track this yourself, you can check with a CO_2 purity tester. This instrument consists of a caustic reservoir and calibrated absorption burette mounted to a PVC frame. The chamber is filled with CO_2 and 20% caustic solution is then added, which reacts with the CO_2 and the purity can be read by the graduations on the burette.

HOW TO MEASURE

There a several instruments available for measuring CO_2 and they range in affordability. An analog device is a great place to start and good to keep around if your digital instruments go down. You will have to pay for precision and accuracy, but it is well worth it if you are sending beer out of house. Knowing precise CO_2 volumes, you can troubleshoot pouring issues and identify re-fermentation issues. There are invasive and non-invasive methods and some meters will read a range of dissolved gases, including dissolved oxygen (O_2) and dissolved nitrogen (N_2).

Volume Meter

Volume meters are fairly simple machines, examples of which are the Zahm and Nagel series #1000D and #6000 and the Foxx Equipment Taprite. Essentially, they are pressure vessels that attach to a tank or package.

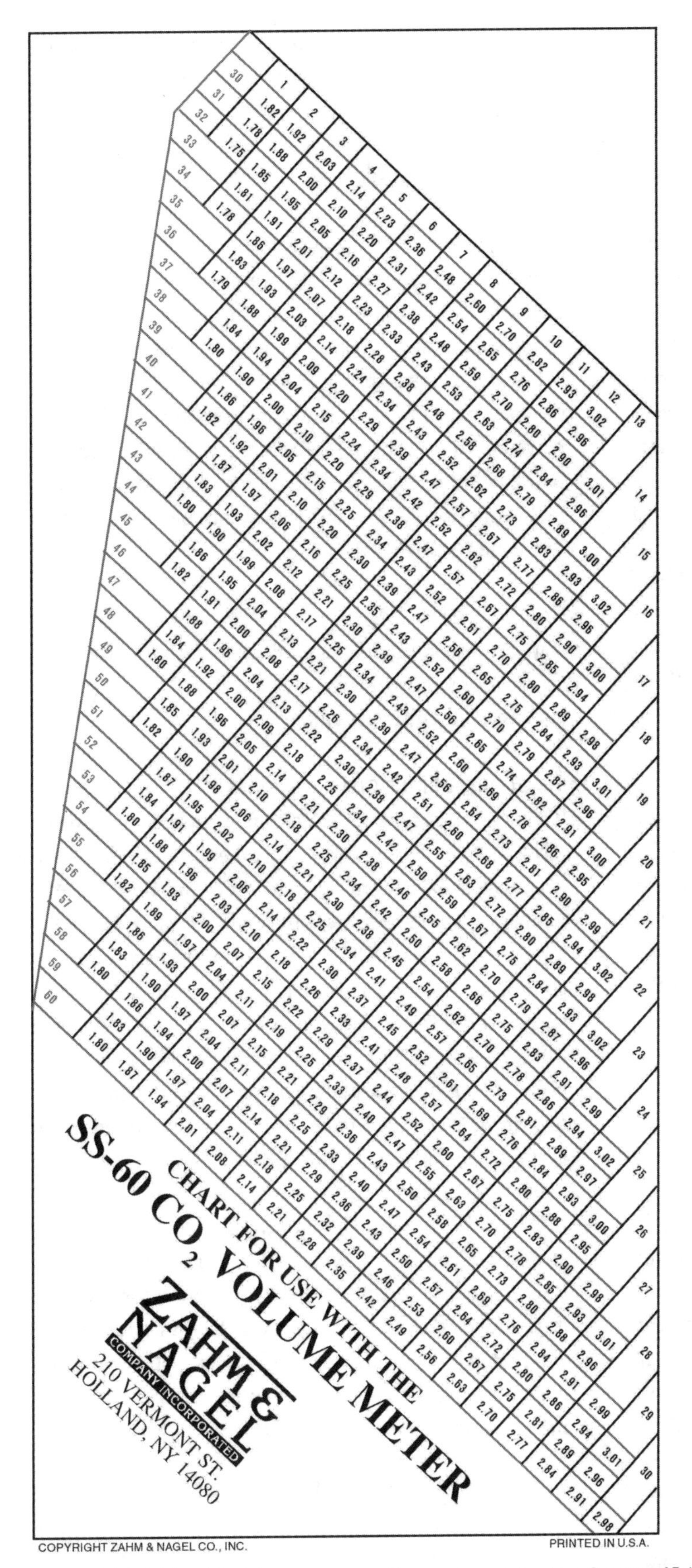

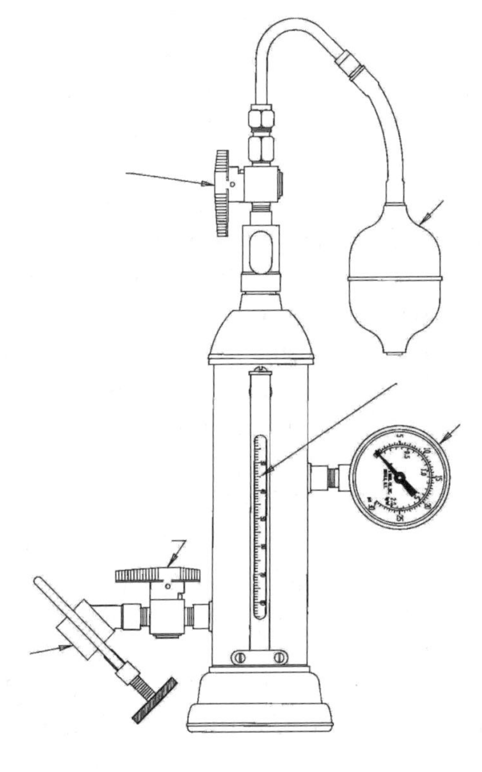

FOR OPERATING INSTRUCTIONS CONSULT THE QUALITY CONTROL TESTING INSTRUMENTS MANUAL.

HIGHER TEMPERATURE AND PRESSURE READINGS ARE SHOWN ON THE CHART IN THE QUALITY CONTROL MANUAL.

Figure 16.1. Solubility of CO_2 in beer–pressure and temperature relationships. Courtesy of Zahm-Nagel.

Beer is allowed to run through the vessel chamber to equilibrate at temperature. Once temperature equilibrated, the vessel is sealed and shaken vigorously. This will equilibrate the gases and put pressure on the gauge. The operator then reads pressure and temperature from gauges attached to the pressure vessel and checks them against a reference, such as the *ASBC Solubility of Carbon Dioxide in Beer Pressure and Temperature Relationships Chart.*

Care must be taken that the pressure and temperature gauges on the volume meter are accurate. If this is the main method of carbonation testing in your brewery, it is a good idea to invest in a "deadweight" pneumatic gauge tester (or something similar) to ensure the accuracy of the pressure gauges. If cleaned regularly and kept well these meters will last a long time.

Digital Meter: Fixed Volume

There are precision instruments that use digital pressure and temperature sensors with a fixed volume chamber. Like the analog volume meter, the instrument is attached to the tank or filling device for packages. Beer is allowed to flow through until foam is eliminated from the sample chamber. The chamber is closed and built-in sensors measure the temperature and pressure of the sample. The instrument then calculates CO_2 volumes using Henry's law. Digital fixed volume meters are precision instruments and should be treated with care to obtain the most reliable results possible.

Digital Meter: Volume Expansion

Other digital meters for measuring CO_2 take advantage of multiple chamber sizes to determine the level of CO_2 and N_2 in the sample. The measuring chamber of the instrument is completely filled with sample and sealed. The device then mechanically expands the chamber to the first set volume and the pressure and temperature is allowed to equilibrate. The equilibrium (absolute) pressure and temperature are then measured. The machine will then mechanically expand the chamber to a second fixed volume. After the second volume expansion, equilibrium pressure and temperature are measured again. From the two data sets of volume expansion and equilibrium pressure and temperature, the CO_2 and N_2 contents are calculated. Volume expansion digital meters are precision instruments and should be treated with care to obtain the most reliable results possible.[2]

Spectrophotometry

The CO_2 content in package can be determined non-invasively by spectrophotometry, that is, passing a particular wavelength of light (a laser) through the headspace of a bottle. Any CO_2 molecules in the headspace absorb the infrared light of the laser. Based on the width of the absorption lines and their intensity, the total pressure and the partial pressure of CO_2 in the package are determined and CO_2 volumes calculated. Using spectrophotometry this way allows for an unlimited number of CO_2 measurements to be conducted on the same package.[3]

SAMPLING PLAN

Carbon dioxide is used all over the brewery for many applications, but measuring CO_2 dissolved in beer is your main focus in the quality lab.

Finished Beer before Carbonation

If you decide to bottle condition your beer, measuring residual CO_2 in solution is very helpful. Residual CO_2 in finished beer can vary greatly depending on the temperature and pressure it is stored at and for how long. Directly measuring CO_2 at this stage means you can be confident in calculating the correct amount of priming sugar to achieve your desired CO_2 volumes in the finished product. As the CO_2 volumes may be very low at this point of your process, make sure you know the limitations of your instrumentation because not all instruments can accurately measure such low levels of CO_2.

Finished Beer after Carbonation

The dissolved CO_2 level after carbonation is an immensely important QCP measurement for the quality minded brewer. Before packaging your beer, ensure you have a true-to-brand specification for CO_2 volumes as it is an important component in the perception of flavor in the finished product. Going beyond flavor,

2 ASBC, "Dissolved Carbon Dioxide," *Methods of Analysis, Beer Methods Beer - 13,* 8th ed. (St. Paul, MN: American Society of Brewing Chemists, 1992).

3 Roland Folz and Frank Verkoelen, "Non-invasive, Selective Measurement for CO_2 in Package Expands Brewers' Quality Control Toolbox" (poster presentation at the World Brewing Congress, Denver, CO, August 13-17, 2016).

TABLE 16.1 CARBON DIOXIDE SAMPLING PLAN

QCP	When and how often	What you are looking for
Brite tank	Before packaging	Appropriate carbonation for brand specifications
Kegs, can, bottle	After startup and at timed intervals	Appropriate carbonation for brand specifications
After packing day	Scheduled intervals	Increase or decrease of CO_2

this after-packaging QCP will serve as a snapshot in time, allowing the quality lab to analyze the condition of packaged beer over time.

After Packing Day

Using your finished and packaged measurements as the baseline, you can check in on the condition of your beer as it ages. Any increase in dissolved CO_2 at this point is a sign that there has been in-package fermentation of some kind. The two main reasons that this occurs is either true terminal gravity was not reached before packing or a more attenuative organism was included in package.

A significant drop in CO_2 can be an indication of a poor package seal. Being proactive about taking measurements post-packing can put you ahead of consumer complaints, hopefully allowing you to catch a quality issue before it becomes a serious issue that will trigger a recall event.

TROUBLESHOOTING

As described in the "How to Measure" section above, each instrument works on slightly different principles. You should consult the manufacturer's operations manual for help with instrument related issues. A few general tips are:

- ensure your measuring instrument is clean and calibrated according to the manufacturer's instructions;
- if results are varied, work on standardizing the measurement with an SOP; and
- check for CO_2 break out, because if gas is escaping from solution you will get an erroneous result.

17 DISSOLVED OXYGEN

Dissolved oxygen is the amount of gaseous oxygen (O_2) present in a liquid. With an ever-increasing number of craft brewers packaging their beer, dissolved oxygen (DO) is a brewery hot topic, and for good reason—O_2 is highly reactive and DO can single-handedly transform your bright and beautiful pale ale into a discolored, stale mess. On the flip side, it is well known that a certain level of DO is desired at pitching to aid in the production of fatty acids and sterols, which are important components of yeast cell membranes.

Oxygen can enter your beer by direct absorption from the atmosphere when sloshing or moving the beer around. Temperature also has an effect, as O_2 more readily dissolves in colder liquids. Keeping O_2 out of your beer post-fermentation is of upmost importance for the craft brewer because it will kill your beer. I have read countless threads about a brewer's beautiful golden IPA turning into a brown muddy mess in a matter of 48 hours. Excess O_2 ingress promotes chemical oxidation reactions that produce aldehydes responsible for undesirable off-flavors, such as papery (*trans*-2-nonenal), cardboard, and sherry-like notes, and can significantly darken the color of your beer. A prematurely oxidized beer is not what your consumer is looking for.

HOW TO MEASURE

Electrochemical DO Sensor

An electrochemical DO sensor typically consists of a probe and a meter to read an electrical signal. The probe has an oxygen-permeable membrane that allows O_2 to diffuse from the sample and react with the cathode inside. The reaction initiates an electrical signal that is proportional to the concentration of O_2 and is read by the meter. These sensors are typically affordable and rugged instruments meant to read in the parts per million (ppm) range, good for determining the aeration of wort.

Optical DO Sensor

An optical DO sensor uses light to measure the DO in a sample. These sensors contain an LED that emits blue light and illuminates an oxygen-sensitive material. Blue light is shone on the oxygen-sensitive material, which emits a red light back that is picked up by a photosensor that measures the intensity and time delay. When O_2 is introduced to the sample stream the intensity of the returning red light is quenched and delayed. This change is detected by the photosensor and is directly correlated to the amount of O_2 present in the sample. Optical DO sensors are highly sensitive and well suited for reading DO in the parts per billion (ppb) range, great for DO measurements post-fermentation.

WHERE TO MEASURE

Wort

Virtually all dissolved gases are driven off from the hopped wort during the boil, including oxygen. Since oxygen is desired for lipid production by yeast, most brewers cool the boiled wort to pitching temperature and inject sterile air or pure O_2 in order to reach a desired DO level. This level is typically in the ppm range. Many authors prescribe a DO "requirement" for hopped wort of 1 ppm per degree Plato, but DO requirements are strain-dependent, so use this suggestion as a starting point. I encourage you to track your pitch rate, hopped wort DO (ppm O_2), yeast performance, and final beer over several batches to decide what DO requirement yields your desired outcome.

Cellar

In a small craft brewery cellaring may be very physical by nature. Cellaring will involve many butterfly valves, long lengths of hose, and most likely a pump or two. It is useful to track O_2 ingress in your fermentation vessels after every major cellaring task. What is a major cellaring task, you ask? A major cellaring task is any time you open the tank to make an addition (dry hopping) or push beer through a length of hose. To determine if you can improve a cellaring process, take a baseline DO measurement before the cellar task and compare it to the DO measurement taken after said task. The goal is to introduce as little O_2 as possible every time you perform a cellaring task.

Package

Most brewers take every precaution to ensure the best beer possible, starting with ingredient selection all the way through fermentation and cellaring until it hits the brite tanks. At this point they leave the rest either up to whatever equipment they have and probably the newest members of their team or in the hands of a mobile canning company to make sure the product of their hard work gets into the package safely. But how does the brewer know their beer is safe and sound in the can/bottle? Many flavors that they worked so hard to achieve, such as polyphenolic hop and malt flavor compounds, are extremely susceptible to oxidation. The meticulous craft brewer must take every measure to minimize oxygen contact with the beer on its journey from fermentor to the package. If beer exposed to high levels of oxygen is packaged, irreversible damage will be done to its flavor profile and the shelf life will be significantly shortened.

SAMPLING PLAN

Your sampling plan should follow the life cycle of your beer. Remember that oxidation is a chemical reaction and is continuously happening. Prompt measurement is important for obtaining accurate data.

If you are injecting pure O_2 or sterile air during knockout use a ppm-range (mg/L) DO meter to measure with either an inline sample port or as soon as the brew kettle is empty. This should happen before your yeast has been pitched because it will quickly begin consuming the O_2.

Moving into the cellar, it is useful to check DO in the fermentation vessel post-fermentation and before cold crashing using a ppb-range (µg/L) DO meter. This DO measurement should be very close to zero and serve as a baseline going forward with cellaring and packing. With that being said, zero DO pickup throughout your process is unlikely, but always strive to improve processes to allow as little O_2 ingress as possible.

The next part of your sampling plan is any time directly before and directly after any other major cellaring task, as described in the section above. Poorly purged transfer lines or leaky pump seals can be a major source of O_2 ingress, so making these data collection points is useful to understand DO pickup in the cellar. Before moving finished beer to the brite beer tank (BBT) it is common practice to purge the empty tank of air (which is mainly nitrogen, oxygen, argon, and carbon dioxide) using CO_2 or water. It is good practice to confirm the purging process was successful because the cold temperatures make the beer extremely susceptible to DO pickup during transfer, even with minimal splashing. Using a ppb-range DO meter to confirm most O_2 has been purged before moving your beer is vital to ensure a long shelf life for your product.

Now that you have safely gotten brite beer into the BBT there is one last hurdle to overcome, getting that beer into the package with as little O_2 as possible. This can be difficult because there are many variables stacked against you, such as cold temperatures, a lot of moving parts needed to fill containers, and the fact that your product fill heads and containers are open to the atmosphere throughout the filling process.

There are three types of DO measurements that we can use to help us control O_2 pickup in the packaging process.

1. Raw $O_2(g)$ measurement
2. Equilibrated DO
3. Total packaged oxygen (TPO)

Up until this point we have been discussing dissolved oxygen as the raw value of gaseous O_2 present at the time of measurement. This is the value displayed by your DO meter. Once in the package we can obtain our second type of DO measurement by shaking the bottle or can vigorously for up to 5 min. to obtain the equilibrated DO value, which simply means that the headspace and liquid gases are in equilibrium and can be measured by your meter. Finally, we use the equilibrated DO values to calculate the total packaged oxygen value, which is the third measurement that relates to DO.[1]

Total packaged oxygen (TPO) accounts for all DO in package. There are DO meters that can directly measure O_2 in the headspace and DO in the beer and give you your TPO, but they are typically expensive. Luckily, TPO can be calculated using a DO value from a shaken package and the equation from C. Vilacha and K. Uhlig to determine the partial pressure of DO in beer measured by the DO meter.[2]

$$m_t = X\left[\frac{32 \times 1000 \times HS(4.15 \times 10^{-7}\, T^2 + 2 \times 10^{-4}\, T - 0.0701)}{0.082 \times T \times 1.0332 \times 100} + 1\right] \quad (17.1)$$

m_t = total oxygen in the container in mg/L
X = oxygen content of the beer in mg/L (or ppm dO_2)
T = temperature in Kelvin (K)
HS = headspace volume as % of beer volume

There are several beer TPO calculators available online for free or to purchase. The ultimate goal is to have the lowest TPO possible. With an understanding of the three types of packaged DO measurements, we can put them to use for troubleshooting DO pickup in the filling process. This is part of optimizing and getting the best possible performance from your packaging line. Table 17.1 breaks up DO uptake in a fill into discrete pieces. Being able to visualize where the uptake is coming from can help you formulate a targeted approach to reducing it. Reducing the baseline O_2, filler O_2, or headspace O_2 will contribute to a lower TPO overall.

Dissolved oxygen pickup in packaging: quick calculations	
Baseline O_2	= raw DO of beer post-carbonation (in BBT or just before filler)
Filler O_2 pickup	= raw package DO (unshaken) − baseline O_2
Headspace O_2 pickup (estimate)	= equilibrated DO − filler O_2 pickup
TPO	Calculated based on equilibrated DO measurement (eq. 17.1); accounts for all DO in package

TROUBLESHOOTING

Even without a DO meter, the following troubleshooting principles can help reduce DO pickup. Of course, measuring is always the best course to take because you can confirm that you are not just wasting CO_2.

- Cellar: evacuate hoses and pumps of oxygen with CO_2 or de-aerated water prior to tank hookup.
- BBT: check brite tank DO content prior to filling and slow down CO_2 flow while purging to discourage gas mixing.
- Filler O_2 pickup: purge hose oxygen bubble by lifting hose and releasing bubble; check proper flow of CO_2 in prepackage fill.
- Headspace O_2: look for tight bubbles to just barely breach the top of package; increase foam on beer (FOB) if it does not breach.

[1] Chaz Benedict, "Total Package Oxygen 101 - The Difference Between Dissolved Oxygen and TPO," *Tap Into Hach* (blog), June 20, 2012, https://tapintohach.com/2012/06/20/total-package-oxygen-101-the-difference-between-dissolved-oxygen-and-tpo/.

[2] C. Vilachá and K. Uhlig, "The Measurement of Low Levels of Oxygen in Bottled Beer," *Brauwelt International*, Volume 1, 1985, 70.

TABLE 17.1 DISSOLVED OXYGEN CONTROL POINTS USED FOR TARGETED TROUBLESHOOTING

QCPs	When and how often	What you are looking for	Expected range (general)
After wort aeration	Inline or after oxygenation	Appropriate O_2 level in ppb as you determine	Approximately 1 ppm per °P
Post-fermentation		Baseline for DO tracking throughout cellaring	>10 ppb
After tank addition	After post-fermentation additions such as finings or dry hops, and any subsequent recirculation	Determine pickup from baseline numbers	>30 ppb
Brite tank	After brite tank purge	Purge efficacy	>300 ppb
Transfer	Before, during, and after transfer	Monitor O_2 pickup	>50 ppb
Brite tank	Before line startup	Baseline for package pickup	>50 ppb
Package	At line startup, timed intervals throughout run	Filler pickup, headspace O_2 pickup, TPO (calculated)	Unshaken >50 ppb Shaken > 75 ppb Calculated TPO >150 ppb

18

ALCOHOL

Whether we admit it or not, alcohol is the reason why we are all here. Civilizations and cultures were built around it, and it is a cornerstone of modern style guidelines, affecting the flavor of our beverages as well as the way they make us feel during and after imbibing. As brewers, we are most often concerned with ethyl alcohol (ethanol) fermentation, where the glucose is broken down by yeasts into pyruvate that is subsequently enzymatically converted into alcohol and carbon dioxide. The general equation for this process is

$$\text{glucose} + 2\ \text{ADP} + 2\ \text{phosphate} + 2\ H^+ \longrightarrow 2\ \text{ethanol} + 2\ CO_2 + 2\ \text{ATP} + 2\ H_2O$$

Brewers control the amount of alcohol produced by adjusting parameters such as recipe formulation, total grist weight, mashing schedules, and yeast selection, all of which can affect the amount of sugars consumed by yeast.

Measuring the alcohol content of your beer satisfies two important QA/QC criteria. First is to make sure your product is within your set product specifications. This can be an indication of the efficacy of your mashing process because that affects your real degree of fermentation (RDF). If your RDF is out of control you should look to your mash protocols and yeast pitching rates and health. The second criterion is to ensure compliance with government regulations and ensure proper taxes are paid, which is a TTB requirement of your brewery. The alcohol by volume (ABV) printed on package should be within ±0.3% of the stated ABV. If you are out of these TTB specifications, you can be subject to fines.

Alcohol by volume (ABV) is a volume percent solution (v/v) defined as the number of milliliters (mL) of pure ethanol present in 100 mL (3.4 fl. oz.) of solution at 20°C (68°F). This represents what portion of the total volume of liquid (in our case beer) is alcohol.

FUSEL AND HIGHER ALCOHOLS

Be aware that there are other pathways that yeast can use to ferment sugars. These alternative pathways do not necessarily result in ethanol production. There are as many as forty different fusel alcohols that can be produced by yeast. They can add warming, hot, or solventlike flavors and aromas to your beer. Providing an optimal environment for yeast fermentation and controlling fermentation conditions to promote cell growth, such as temperature, aeration, and free amino nitrogen, can reduce the formation of fusel alcohols.

Alcohol by weight (ABW) is a weight percent solution (w/w) defined as the weight of pure ethanol present in the total weight of solution. The density of ethanol is 0.78945 g/cm^3 at 20°C (68°F), which is less than the density of water at the same temperature and volume, 0.9982336 g/cm^3. Because of this fact, ABW is always lower than ABV

HOW TO MEASURE

Calculating ABV by Tracking Fermentation

Determining the alcohol content of your beer can be done principally by tracking fermentation and using the original extract and final apparent extract values to calculate the ABV. This is an acceptable method but it is only as accurate as your extract measurements. Depending on your instrumentation and skill, this may or may not report your ABV within governmental regulations. Regardless, this QCP should be monitored every brew to ensure you are meeting your product specifications. Before we work through an example, you should refamiliarize yourself with the terms in table 18.1 that were first introduced in chapter 9.

Example ABV calculation

Let us put those terms in table 18.1 to work and go through an ABV calculation. We will use values of original extract (OE) and final apparent extract (AE) and also refer to the mass balance for the equation for the fermentation of sugar to ethanol.

After brewing a pale ale, suppose you measure the OE to be 12.5°P (1.050 SG) and AE at the end of fermentation to be 2.5°P (1.010 SG). To get to ABV, first use these fermentation measurements to determine real extract (RE).

$$\begin{aligned} RE &= (0.1886 \times OE) + (0.8114 \times AE) \\ &= (0.1886 \times 12.5°P) + (0.8114 \times 2.5°P) \\ &= 4.4°P \end{aligned}$$

With your RE value, you can then determine your RDF.

$$\begin{aligned} RDF &= 100 \times \left(\frac{OE - RE}{OE}\right)\left(\frac{1}{1 - (0.005161 \times RE)}\right) \\ &= 100 \times \left(\frac{12.5°P - 4.4°P}{12.5°P}\right)\left(\frac{1}{1 - 0.005161 \times 4.4°P}\right) \\ &= 66.4\% \end{aligned}$$

TABLE 18.1 OVERVIEW OF DENSITY RELATED TERMINOLOGY

Parameter	Description	Equation
Original extract (OE) (a.k.a. original gravity, OG)	Solids extracted from grist as % wt./wt. It is convenient to use degrees Plato for these equations	$\left(\frac{\text{weight extract}}{\text{total weight wort}}\right) \times 100 = °P$
Apparent extract (AE)	After fermentation has started, AE is the direct measurement of total extract in wort or beer not corrected for alcohol content. Alcohol has a specific gravity substantially lower than water. This means the "true" or real extract (RE) is greater than that measured directly	Direct measure of fermenting wort or beer
Apparent degree of fermentation (ADF) or apparent attenuation	Observed reduction of wort extract not accounting for the density of alcohol in solution	$ADF = 100 \times \left(\frac{OE - AE}{OE}\right)$ AE reading taken at end of fermentation
Real extract (RE)	Total extract in wort corrected for the actual amount of alcohol in the wort. RE calculations account for the presence of alcohol in the finished beer and the absence of alcohol in the starting wort	$RE = (0.1886 \times OE) + (0.8114 \times AE)$ Use °P for OE and AE
Real degree of fermentation (RDF)	RDF is the measured percentage of wort extract that is fermented	$RDF = 100 \left(\frac{OE - RE}{OE}\right)\left(\frac{1}{1 - (0.005161 \times RE)}\right)$

Now that you have the RDF, you can determine what proportion of your OE was water, sugar, and unfermentable dextrin. Looking back at chapter 9, we know that 1°P is a 1% sugar w/w solution (at 68°F). Therefore, 100 g of 12.5°P wort comprises 87.5 g H_2O and 12.5 g extract. Your RDF tells you the 12.5 g extract was only 66.4% fermentable sugars:

$$12.5 \text{ g extract} \times 0.664 = 8.3 \text{ g fermentable sugar}$$

This means that

$$12.5 \text{ g extract} - 8.3 \text{ g fermentable sugar} = 4.2 \text{ g dextrin}$$

You now know that 100 g of the measured wort has 87.5 g water, 8.3 g fermentable sugar, and 4.2 g dextrin.

Let us look again at the fermentation mass balance equation first introduced in relation to CO_2 production in chapter 16:

$$\begin{array}{ccccccccc} C_6H_{12}O_6 & = & 2\,C_2H_5OH & + & 2\,CO_2 & + & \text{yeast growth} & + & \text{by-products} \\ 100\text{ g} & = & 48.6\text{ g} & + & 46.4\text{ g} & + & \approx 3\text{ g} & + & \approx 2\text{ g} \end{array}$$

This is only applied to the 8.3 g sugar, since dextrin is not fermentable. The equation shows that 48.6% of the sugar is made into ethanol, 46.4% into CO_2, and the remaining 5% is used up in yeast growth and the by-product formation.

$$\begin{aligned} 8.3 \text{ g sugar} \times 0.486 &= 4 \text{ g ethanol} \\ 8.3 \text{ g sugar} \times 0.464 &= 3.9 \text{ g } CO_2 \\ 8.3 \text{ g sugar} \times 0.05 &= 0.4 \text{ g yeast + by-products} \end{aligned}$$

In most breweries, the CO_2 is lost to the atmosphere through the blow-off tube and the majority of the yeast and by-products are removed. With those removed the total mass of the beer will be

$$100 \text{ g} - 3.9 \text{ g} - 0.4 \text{ g} = 95.7 \text{ g beer}$$

To determine percentage alcohol wt/wt (i.e., ABW) you divide the mass of alcohol produced by the total mass of beer:

$$\begin{aligned} \text{ABW} &= 100\left(\frac{\text{alcohol produced (g)}}{\text{total mass (g)}}\right) \\ &= 100(4.0 \text{ g}/95.7 \text{ g}) \\ &= 4.2\% \end{aligned}$$

As stated before, the density of ethanol is less than the density of water at the same temperature. We can use this fact to convert ABW to ABV by dividing ABW by the relative density of ethanol and water. This will be the constant, 0.79:

$$\begin{aligned} \text{ABV} &= \frac{\text{ABW}}{\left(\frac{0.78945 \text{ g/cm}^3}{0.9982336 \text{ g/cm}^3}\right)} \\ &= 4.2/0.79 \\ &= 5.3\% \end{aligned}$$

Calculating ABV by Distillation

Finished beer is a complex mixture of water, alcohol, dextrin, and other solids, all with differing densities. Determining the ABV using distillation requires separating the ethanol from this mixture. To achieve this, we take advantage of the fact that ethanol has a lower boiling point (173.3°F, 78.5°C) than water (212°F, 100°C) and the other constituents in beer. The distillation equipment consists of a closed system in which the beer is heated to 173.3°F, evaporating the ethanol and allowing it to rise into a condenser that empties into a collection vessel. Once distillation is complete, water is added to the collected ethanol to create a binary mixture with a final fixed volume (100 mL). The density of the ethanol and water mixture is measured and the ABW or ABV derived using the calculations in *ASBC Methods of Analysis* "Beer-4a" or "Beer-4b," respectively.

Near-Infrared (NIR) Spectrophotometry

A near-infrared (NIR) spectrophotometer determines the ABV (%, v/v) content of a beer using absorbance of light at NIR wavelengths. The density is determined by means of a densitometer. While it is possible to determine alcohol content from NIR spectrophotometry alone, an alcolyzer combines an NIR spectrophotometer and densitometer in one device. These instruments can be expensive new, but they are very accurate and precise. From the absorbance and density values, the instrument calculates a number of beer parameters, including ABW, apparent extract, real extract, original extract, real degree of fermentation, calories, specific gravity, apparent degree of fermentation, present gravity, original gravity, spirit indication, extract gravity, degrees lost, and the extract-to-alcohol ratio.[1] Spectrophotometry measurement principles are covered in closer detail in chapter 20.

WHERE TO MEASURE

Alcohol and extract go hand in hand, but you will want to measure or calculate ABV only after final gravity is achieved. This is a major QCP for brewers. If the expected final gravity is not achieved the QA/QC scientist should figure out if the beer has truly stopped or if it will continue to ferment. More fermentation means more alcohol, and this can drive your product out of specification. If you are indeed at final gravity you must ensure that the beer is within tolerances for government regulation. If you are out of this tolerance you have a decision to make: blend it with a higher/lower ABV product or sterile de-aerated water to get it in range, or dump and start again. Of course, these actions do have consequences pertaining to sensory and other qualities of the product.

SAMPLING PLAN

When it comes to ABV, you are looking to hit two marks. First and foremost is making sure the ABV of your product matches the printed ABV on the package within ±0.3% ABV. Second is making sure the ABV hits your internal product specifications for the brand. If you think about it, these should be the same number. It is unlikely that you will have a internal specification different to the statement you put on the package. If they are different it is time to align them, as it is not easy to hit two targets with the same arrow.

TROUBLESHOOTING

There are a few common trouble areas where ABV consistency can be affected.

Mash tun:

- I have found the biggest culprit in our brewery affecting ABV is missing grist in the mash. We have employed a triple-check policy to ensure all grain gets weighed and milled correctly.
- Ensure mash temperatures and pH are in targeted ranges. Being above or below your mash temperature and pH will favor a different set of enzymes, which affect conversion of sugars that will in turn influence alcohol production.

Yeast management:

- Ensure healthy viable yeast is used at a consistent pitch rate.

TABLE 18.2 ALCOHOL SAMPLE PLAN

QCPs	Frequency	What we're looking for
At end of fermentation, after terminal gravity is reached before packaging	Each fermentation	ABV within product standards and TTB label specification
After blending project	Each blend	ABV within product standards and TTB label specification
After barrel aging in spirit barrel	Each blend	ABV within product standards and TTB label specification

1 "Beer Methods Beer - 4, 'Alcohol'," in *Methods of Analysis*, 8th ed. (St. Paul, MN: American Society of Brewing Chemists, 2001).

19

TITRATIONS

Titration is a useful method of chemical analysis to determine an unknown concentration of an acid or a base. To understand how this works, we will draw on some concepts already discussed in chapter 11. The first is the pH scale. As we saw before, an aqueous acidic solution (pH zero to 6.99) is where the hydrogen ion (H^+) activity is greater than hydroxide ion (OH^-) activity, and vice versa for an aqueous basic solution (pH 7.01–14.0). Notice the charges on the two ion species. Theoretically, when we mix acids and bases the opposite charges of H^+ and OH^- attract and they chemically combine to form H_2O, therefore neutralizing each other. When a strong base is mixed with a strong acid this neutralization reaction results in a pH of 7. When a weak base is mixed with a strong acid the result is a pH <7; conversely, a strong base mixed with a weak acid results in a pH >7.

A titration involves slowly adding a standardized solution of known concentration (the **titrant**) to a specific volume of unknown solution (the **analyte**). This forces the neutralization reaction, because the concentration of H^+ ions will eventually exactly equal the concentration of OH^- ions in the solution so that their activity is neutralized. This ideal point where the neutralization reaction goes to completion is called the **equivalence point**. The equivalence point is usually revealed by a color change indicator or a specific pH. Because the ions of our known titrant are exactly equal to the ions in the analyte when the equivalence point is reached, you can determine the concentration of the unknown acid or base.

In the brewery we work with acids and bases all the time. Being able to determine concentrations via titrations opens up a world of possibilities. You can use titrations to verify cleaning and sanitizing chemicals are at an effective level, thus minimizing the risk of microbiological infections. This can also lead to saving money by not over-using cleaning and sanitizing chemicals. You can also use titrations to dig deep into your water chemistry and take control of your mash pH. Titrations can also be used to quantify the concentration of organic acids in beer or wort, which is referred to as **titratable acidity**. Measuring the titratable acidity of your beer or wort is the first step to creating a consistently sour product.

SETTING UP FOR A TITRATION

Let us look at what you need to set up a traditional titration to find the equivalence point for a chemical of unknown concentration. If you are performing a titration following ASBC Beer-8, figure 19.1 shows what your set-up will look like. This may cause you to flash back to chemistry class for a minute, but that is really it. Simple and powerful.

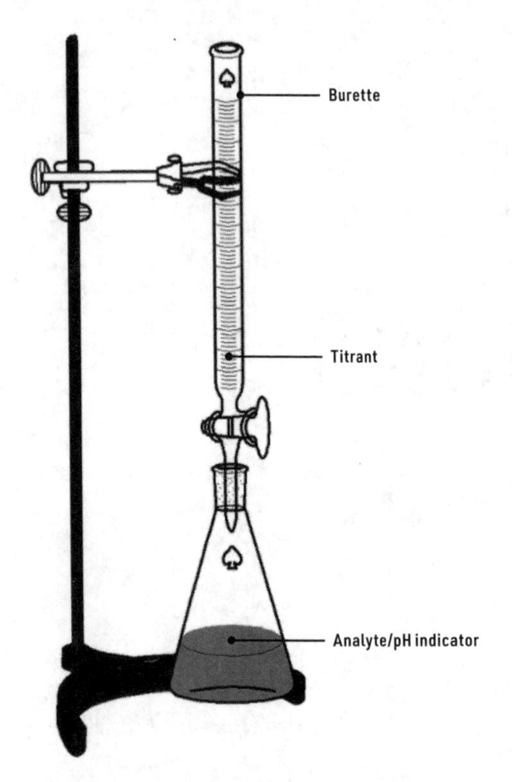

Figure 19.1. Components of a titration.

- **Titrant** – standardized solution (i.e., known concentration) of acid or base to be added to the analyte
- **Analyte** – sample solution of unknown concentration to be titrated
- **Burette** – a graduated glass tube with a tap at one end for delivering known volumes of a liquid
- **pH indicator** – a compound added to the analyte that changes color in a solution over a narrow range of pH values. Used as a visual indicator of the equivalence point. A pH meter could also be used here.

TITRATION OF A STRONG ACID WITH A STRONG BASE

For a titration to work the titrant and analyte must have opposite acid-base properties. In other words, one must be an acid and the other a base. For the titrant, it is advantageous to use a strong acid or strong base because it will fully disassociate its ion, either H^+ or OH^-, respectively. Earlier I described the concept of neutralization in terms of the ionic exchange between a strong base and a strong acid, but I left out some details. Let us look closer at the titration of a strong base and strong acid to get an idea of what is going on here.

Suppose we have 50 mL of hydrochloric acid (HCl) of unknown concentration, which is our analyte. In order to determine the concentration, we decide to titrate with a standard solution (see chap. 8) of 0.10 M NaOH, which is our titrant. The neutralization reaction for this is

$$HCl + NaOH \longrightarrow NaCl + H_2O\,.$$

The reaction products are salt and water. Remember that the strong acid and strong base completely dissociate in solution into their constituent ions, so we can write this equation in terms of the ions involved:

$$H^+ + Cl^- + Na^+ + OH^- \longrightarrow Cl^- + Na^+ + H_2O$$

As far as the pH of the solution is concerned, Cl^- and Na^+ do nothing to affect the pH, so the net ionic equation involved in the above reaction is simply

$$H^+ + OH^- \longrightarrow H_2O\,.$$

Measuring the pH of the analyte against the amount of NaOH added shows the course of the titration (fig. 19.2). As we slowly add NaOH the pH slowly rises as the additional OH^- ions react with the excess H^+ to form water. This continues until we come close to the equivalence point, when the pH increases rapidly. The steepest point of this curve is the equivalence point and happens to be pH 7. At the equivalence point we read on the burette that we have added 50.0 mL of titrant. (As we go past the equivalence point we see the solution rapidly becomes basic because the HCl is completely consumed and we are adding excess OH^-.) Now that we know the amount of titrant added to reach the equivalence point, we can solve for our unknown concentration of analyte.

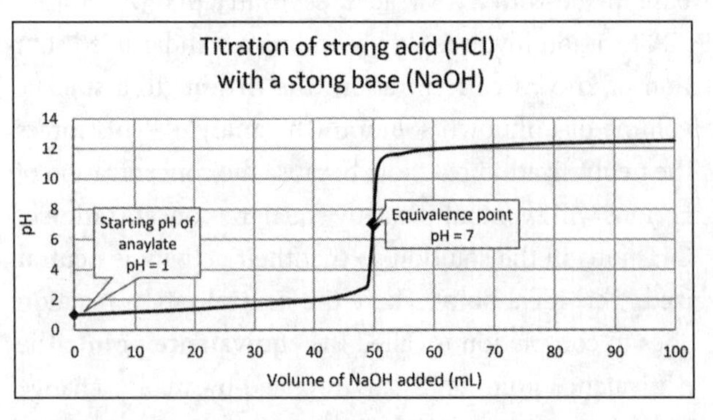

Figure 19.2. Titration curve for titration of HCl with 0.10 M NaOH standard solution.

First multiply titrant volume by concentration. Remember that the units of molarity (M) are moles per liter, so the titrant volume must be expressed in liters. In this example, we added 50 mL of titrant, which is 0.05 L.

$$0.05 \text{ L} \times 0.10 \text{ M NaOH} = 0.005 \text{ mol NaOH}$$

This means 0.005 moles of titrant was added to neutralize the analyte and reach the equivalence point. From the net ionic reaction above, this means the analyte contains 0.005 moles of HCl, because we needed to add 0.005 mol OH^- to neutralize all of the excess H^+.

To obtain the molarity of the analyte we simply divide the the starting volume of the analyte sample by the number of moles we now know the sample contained. In this case, we started with 50 mL (0.05 L) of HCl:

$$0.005 \text{ moles}/0.05 \text{ L} = 0.10 \text{ M HCl}$$

This result makes sense because we are titrating a strong acid with a strong base. We know that each completely dissociates into H^+ and OH^- and we also know that the equivalence point will be neutral, that is, pH 7.

I should point out that you will not work exclusively with strong acids and bases in the brewery. While the set-up and procedure for titration is the same, if the analyte being titrated is a weak acid or weak base the equivalence point will not equal pH 7. Each acid-base pair will have a unique pH at its equivalence point.

- A weak acid titrated with a strong base leads to a pH >7 at the equivalence point.
- A weak base titrated with a strong acid leads to a pH <7 at the equivalence point.

HOW TO MEASURE

There are two main ways that we perform titrations in the brewery lab. The first is with a burette and beaker using pH or a color changing indicator to determine the end point. This is the method used in the example from the previous section. The second method is a drop titration kit, which you should able to source from your chemical supplier.

Burette and Beaker

This method requires a bit more planning and patience than the drop titration kits, but results can be much more accurate due to the ability to titrate at a larger sample size. You can also measure just about any chemical if you dig in and do some research on acid-base equilibria, equivalence points, and finding the appropriate titrant and indicator. We have just touched the surface in this chapter.

Auto-Titrator

This still uses a burette and beaker but, as the name suggests, an auto-titrator performs the titration for you. The instrument consists of a piston driven burette for accurately and precisely dosing titrant into a titration beaker filled with analyte. The titrant dosing is based on readings from a built-in pH electrode located in the titration beaker. Using an auto-titrator can be fast and accurate, especially for brewers who focus on sour beers. Many other industries use auto-titrators, so they come in many shapes and sizes and perform specific titration methods for each analyte being measured. Be sure to purchase a titrator that can perform (or be programed to) the titration you want, such as titratable acidity or total alkalinity titrations.

Drop Count Titration Kits

Drop titration kits work on the exact same principle as the classic titration method, except no burette is required, you use a dropper supplied with the titrant. The kits come with illustrated instructions and are typically quick and easy to use. Slowly drop chemical straight from the bottle, count each drop, and, when the desired color change is achieved, simply multiply by the conversion factor provided to determine the concentration of analyte (active product).

Often chemicals from your supplier are formulated products and contain proprietary amounts of the key ingredients, so make sure you have a drop count titration kit for each chemical you use. A good chemical supplier should be able to provide you with a custom titration kit if one does not already exist for the chemical you are interested in measuring.

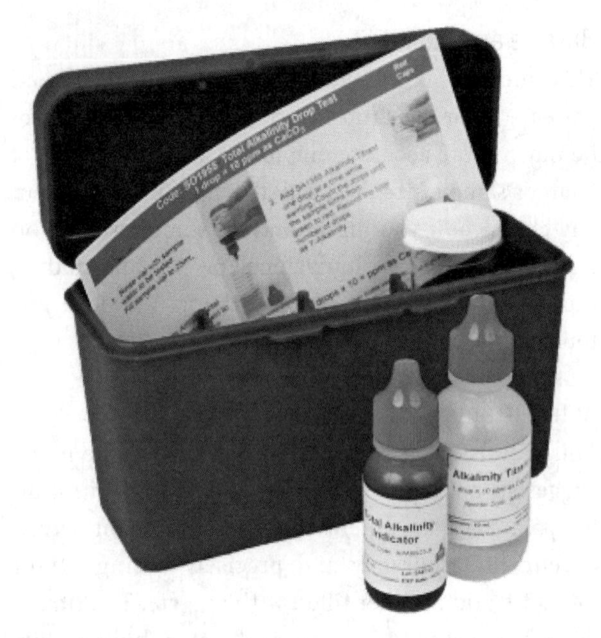

Figure 19.3. A drop titration kit can be used to assess the concentration of your cleaning chemicals.

WHERE TO MEASURE

Phenolphthalein and Total Alkalinity Titrations

The alkalinity of water is primarily due to the presence of hydroxide (OH^-), carbonate (CO_3^{2-}), and bicarbonate (HCO_3^-) ions. Phenolphthalein is a pH indicator that can be used for titration with an acid down to pH 8.3, this endpoint being a measure of phenolphthalein alkalinity. Total alkalinity can be determined by titrating with an acid and measuring the pH to the appropriate equivalence point. When the phenolphthalein alkalinity and total alkalinity are known, the concentration of OH^-, CO_3^{2-}, and HCO_3^- in a sample can be determined. This is the place to start when you are trying to control your mash pH.

Titration of Cleaning Chemicals

Before each CIP task, you can use titration to check you are dosing chemicals to the manufacturer's recommendations. A drop count titration kit is a great way to do this. Typically, your chemical supplier will supply you with a kit free of charge for each chemical you purchase for the brewery. In addition, you can titrate after CIP to make sure your chemical was not neutralized by constituents in the tank and stayed in the working range. These titrations are the basis of auditing and optimizing your CIP process.

Titratable Acidity in Beer or Wort

Organic acids play an important role in the flavor of beer. The level of organic acids in your beer can be determined by measuring titratable acidity (TA), which is defined as the number of H^+ that the organic acids present can potentially donate. When establishing a brand of sour beer, TA is a more suitable measure of sour flavor than pH alone. pH only measures the activity of free H^+ ions in solution, whereas TA will measure free H^+ as well as those bound to any organic acids that are present. This means TA can tell you the total acid concentration of your beer, known as total acidity, which can inform you about the flavor impact more than a simple pH measure can. Refer to the ASBC *Methods of Analysis* "Wort – 7, 'Total Acidity'" or "Beer – 8, 'Total Acidity'" for detailed instructions.

SAMPLING PLAN

QCPs	When and how often	What you are looking for
Cleaning chemicals	Every clean until dosing procedure is validated	Appropriate concentration recommended by chemical supplier
Phenolphthalein and total alkalinity titration	Schedule based on your water supply	Water alkalinity
Sour fermentation	After souring process	Titratable acidity (ASBC method total acidity) in mol/L
Packaging a sour beer	Every sour beer pack	
Kettle sour wort	Use titratable acidity as a marker to end a kettle sour	

TROUBLESHOOTING

Be sure to standardize your acid or base titrant (see chapter 8 "Making Standard Solutions" p. 73). Alternatively, compare your standard with a standard provided by a chemical supply company.

If using pH as an indicator of equilibrium try a three-point calibration of the pH meter at pH 4, 7, and 10 for best results.

Titrations are a powerful measuring tool in the hands of a brewer. Using titration, unknown concentrations of chemical compounds can be determined with little effort. It can be used to optimize cleaning processes, saving you time and money in the brewery. Using titration in the production or blending of sour beer will help with brand consistency from batch to batch. Understanding and implementing QCPs around titrations will take the guesswork out of many chemical-driven processes around your brewery.

20

SPECTROPHOTOMETRY

Spectrophotometry has its roots in spectroscopy, the study of the interaction between matter and electromagnetic radiation. It allows the brewing scientist to quantitively measure the extent of light absorption and relate it to the concentration of solute in the sample liquid.[1] Every substance will transmit or absorb certain wavelengths of light based on its unique molecular makeup. A spectrophotometer measures the peaks in absorption and, with a little help from math, you can calculate the molar quantity of the substance present. While spectrophotometry may not be the first set of measurements that you include in your quality program it is a powerful tool for the brewing scientist and can take you to the next level of refinement in you QA/QC program.

PRINCIPLES OF MEASUREMENT

A spectrophotometer consists of a light source that is split into its component wavelengths by a prism or diffraction grating. The light is shone onto your liquid sample, which is held in a cuvette of fixed length. The light that passes through the sample then reaches a detector. The spectrophotometer "scans" all wavelengths of light transmitted to the detector and determines the amount of light absorbed at each wavelength by the sample. The result is a plot of light absorbed as a function of wavelength.

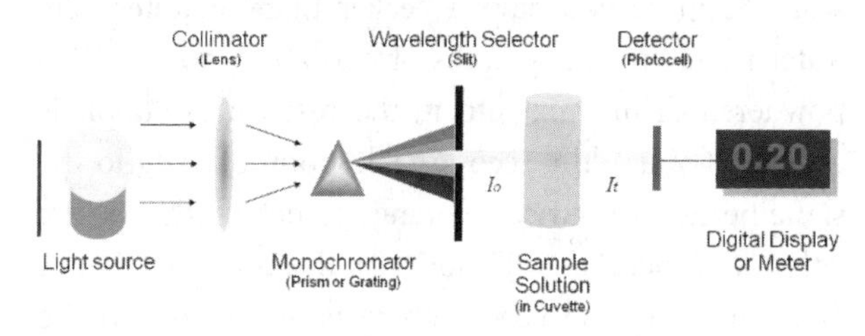

Figure 20.1 Basic structure of spectrophotometers (illustrated by Heesung Shim); in open source https://chem.libretexts.org/Core/Physical_and_Theoretical_Chemistry/Kinetics/Reaction_Rates/Experimental_Determination_of_Kinetcs/Spectrophotometry.

That is all well and good, but how do we measure concentration using the results of the spectrophotometer? First, we need to understand transmittance, absorbance, their relationship, and the aptly named Beer–Lambert law. **Transmittance** (T) is the ratio of light passing through the sample (transmitted light, I_t) relative to the initial light shone onto the sample (incident light, I_0).

$$\text{Transmittance} = \frac{\text{intensity of transmitted light}}{\text{intensity of incident light}}$$

$$T = \frac{I_t}{I_0}$$

[1] John C. Kotz et al., *Chemistry & Chemical Reactivity*, 8th ed. (Belmont, CA: Brooks/Cole, 2009), 188.

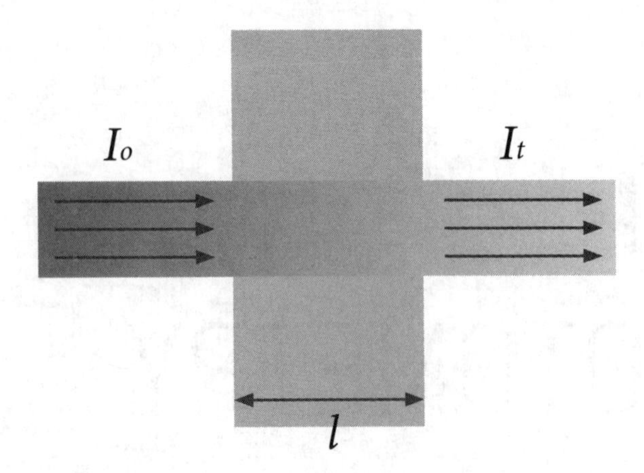

Figure 20.2. Illustration of transmittance, I_t, and path length, l.

Absorbance and transmittance have an inverse relationship. As transmittance increases absorbance will decrease. **Absorbance** (A) is, therefore, defined as the negative logrithm of its transmittance.

$$\text{Absorbance } (A) = -\log(T)$$
$$= -\log\left(\frac{I_t}{I_0}\right)$$

Let us put this together with an example on the macro scale. Suppose you have a beaker of pure water. The water is colorless. If you dissolve a tablespoon of Tang powdered drink mix into it, the resulting solution is orange. If you add a second tablespoon of Tang to the same beaker it is twice as orange as before, that is, the color is deeper. You will notice as the concentration of Tang increases the more light gets absorbed and the less light gets transmitted through. This makes sense, *as the concentration increases, so does the absorbance of light.* This simple example is how we are going to relate absorbance to the chemical concentration of substances in our beer. There is just one more factor to consider, the length the light must travel. If we take our Tang solution and pour it into a shot glass and a pint glass, you will see the liquid in the shot glass appears lighter than the pint, even though both vessels contain the same concentration solution. That is because the transmitted light from the pint has traveled a greater distance through the solution. This also makes sense—if you have ever been deep-sea diving or watched a nature documentary about the ocean, you will have seen the further you dive the more light gets absorbed and the darker it gets as you go deeper, so much so that much of the ocean floor is pitch dark. We can now say that *absorbance increases as path length increases.*

Putting these two facts about concentration and path length together, we can construct the Beer–Lambert law. Written mathematically, absorbance (A) is proportional to the path length (l) times concentration (c):

$$A \propto l \times c$$

Without getting too mathematical, we will introduce the proportionality constant (ε), also called molar absorptivity, to transform our proportionality to an equation expressing equality:

$$A = \varepsilon \times l \times c \qquad \text{(Beer–Lambert law)}$$

where

- A = the absorbance of the sample (dimensionless)
- ε = molar absorptivity (L/mol·cm) given constant pressure, temperature, and wavelength
- l = length (cm)
- c = concentration (mol/L)

This is a powerful equation for measuring a large variety of relatively dilute chemical solutions. Note that at higher concentrations the dependence of absorbance on concentration may not be linear.

TURBIDITY

Turbidity refers to the cloudiness of a fluid, otherwise known as haze. It is caused by large numbers of individual particles suspended in solution. The suspended particles block, scatter, and absorb light, making the beer appear cloudy, hazy, or turbid in varying degrees. Haze can result from many causes: proteins, polyphenols, yeast cells, starch, bacterial infection, chemical inclusion (residual cleaning or coolant compounds), and even simple carbohydrates in colloidal form. The interaction of suspended particles and light does not follow the Beer-Lambert law and, therefore, it should not be measured using a spectrophotometer. To measure turbidity, you need a turbidimeter or a nephelometer, which use scattered light (not the absorption of light) to measure turbidity.

HOW TO MEASURE

Depending on the substance you want to measure you require a specific spectrophotometer for the wavelength range you are working in:

- **UV-visible spectrophotometer:** uses light over the ultraviolet range (185–400 nm) and visible range (400–700 nm).
- **IR spectrophotometer:** uses light over the infrared range (700–15,000 nm).

Your sample will also need to be prepared in a way that depends on the parameter you are measuring, be it color, FAN, alpha acids, etc. In general, spectrophotometric analysis can be carried out using the following general procedure.

First, obtain the absorption spectrum for the substance being analyzed by scanning all wavelengths in the device's range and looking for peaks in absorption. Since absorption at each wavelength is proportional to concentration, you can theoretically analyze a sample using just about any wavelength, but at maximum absorption the change in absorption is greatest for a given change in concentration. So, to obtain the most precise measurement possible you should generally choose wavelengths of maximum absorbance for your measurements. Most beer methods you will use already have this peak wavelength determined for you.

Next, make a calibration plot at the chosen wavelength for the substance. This is done by preparing a standard solution (see chapter 8) of accurately known concentrations and plotting its absorbance on the y-axis against concentration on the x-axis. As long as the solution is dilute and follows the Beer–Lambert law, absorbance and concentration will have a linear relationship and should produce a straight line with a positive slope. Using spreadsheet software, you can enter this calibration data and draw a calibration plot and perform linear regression to obtain the equation for the line.

Finally, measure the absorbance of the sample you wish to analyze. Plug in the measured absorbance for y in your linear regression equation and use that to solve for x. The value obtained for x is the concentration of your sample.

The good news is most modern spectrophotometers will do a lot of this work for you. In fact, many come preprogramed with *ASBC Methods of Analysis* methods for your use.[2]

WHERE TO MEASURE

Think carefully about which parameters you want to analyze using spectrophotometry. Some methods are more complex than others, requiring a range of sample preparations. Some methods only require de-gassing the sample, while others require distilling the beer and adding reagents. All the listed methods in *ASBC Methods* are useful but trying to keep up and track every single parameter requires a large amount of time and resources. Stick with a few methods that are important to your process. For a sub-10,000 bbl. craft brewer, color, beer bitterness, and VDKs (diacetyl and 2,3-pentanedione) will likely be the most useful. However, bear in mind that VDK analysis by spectrophotometry is a two-step process requiring distillation of decarbonated beer and preparation of the sample using a color-changing reagent before it can be measured in the spectrophotometer.

A spectrophotometer may not be necessary for every sub-10,000 bbl. craft brewer, but I would think about adding a spectrophotometer when upgrading or scaling up your brew system. Not every kettle performs the same, making brands consistency difficult when scaling up or changing equipment. Getting baseline numbers for color and IBU will help with recipe adjustment when the time comes to make physical changes to your brewhouse. When adding a spectrophotometer, make sure you allocate enough time and personnel to track data over a large sample range, especially when considering important brands that are to be scaled up. Plot all relevant data in control charts, determine realistic control limits, and use those to inform adjustments you need to make.

SAMPLING PLAN

Your sampling plan will depend on what parameters you decide are worth measuring spectrophotometrically. The following is a list of useful parameters that can be measured using a UV-visible spectrophotometer. The relevant section of *ASBC Methods* is also given.

---

2 You should reference *ASBC Methods of Analysis* for the parameter you want to measure spectrophotometrically, as it contains useful information regarding calibration and sample preparation. Accessing *ASBC Methods* requires a subscription to the organization but it is well worth it.

- **Color** – ASBC Beer-10
    - Easiest parameter to measure using UV spectrophotometer
    - Can also be measured with a comparator
    - Simple, quick measurement that can show deviations in the brewhouse, e.g., recipe, hot time, hot side oxidation
    - Samples need to be filtered prior to measurement
    - Dark beers will need to be diluted or measured with a smaller cuvette (1 mm instead of 1 cm)
- **Beer Bitterness** – ASBC Beer-23
    - Useful measure for determining IBU
    - UV spectrophotometer must measure accurately at 275 nm
    - Can help a great deal when determining hop addition times during recipe development
    - Consistent bitterness is key to maintaining a consistent flavor profile
    - Uses isooctane (2,2,4-trimethylpentane), an organic solvent that needs to be properly disposed of or regenerated. DO NOT PUT DOWN DRAIN!! Isooctane regeneration saves on both consumables and hazardous waste disposal costs.[3]
- **Diacetyl** – ASBC Beer-25B
- **Iron concentration** – ASBC Beer-18A, C
- **Thiobarbituric acid index** – ASBC Wort-21
- **FAN** – ASBC Wort-12
- **Total polyphenols** – ASBC Beer-35
- **SO_2** – ASBC Beer 21
- **Alpha acids and beta acids in hops** – ASBC Hops-6
- **Hop storage index** – ASBC Hops-12

Before you take the plunge into using spectrophotometric methods of analysis, be sure to have a clearly laid out plan and the lab personnel to make your selected measurements. Determine what methods are most relevant to your process and start with those, setting up internal standards for each brand. Once the personnel and standards are in place, you and your team can work on honing the process to meet those standards every time.

TROUBLESHOOTING

Each parameter and method mentioned in this chapter will present a different set of difficulties, too varied to predict and address here. If you are having trouble, reach out to your local university chemistry or food science department. Establishing a working relationship with a local university can be mutually beneficial to both parties. Professors love to see real-world applications of the scientific principles they are teaching in class and will likely be able to help you out. This relationship can provide you with resources and equipment you may not be able to afford yourself as well as provide real-world experience for their students.

[3] Benjamin T. Bailey, "A review of the cost analysis of iso-octane regeneration for re-use in beer bitterness analysis" (poster presented at MBAA Annual Conference, Austin, TX, October 23-26, 2013).

PART IV
GETTING TO WORK

You are quickly constructing your quality pyramid and learning about, and possibly already implementing, your prerequisite programs (GMP and food safety) that will form the foundation of your quality system. You are writing your quality manual and putting SOPs in place to make it work. You have identified various QCPs for process controls in your quality plan and are now armed with 101 ways to measure them.

Now you are going to learn how to use those measurements and apply a variety of analytical methods to set up tolerances for your standards, control your processes, and consistently brew the best beer you can! Before you can get to work and start analyzing your measurements, however, you must set up your lab space.

21

SETTING UP YOUR BREWERY LAB

The most basic of brewery labs should serve two main functions: provide a workspace for measurements and experiments, and provide a place to log and analyze data. You do not need a vast lab space when starting up. In fact, it may just be a designated area or shared office space. Whatever path you choose for your brewery, a dedicated space is integral to launching a successful program. A dedicated space will allow you to focus on the task at hand, which can typically be a big-picture issue compared to some of the day-to-day tasks of brewing.

Getting started is the hardest part. At small and start-up breweries, space is at a premium. Do not let space stop you from starting a lab or QA/QC system—simply start with what you have. Some of the most successful QA/QC systems I have had the chance to visit started in spaces that were literally just broom closets, such as the labs of Allagash Brewing Company and Alaskan Brewing Company. They have since grown into successful enterprises! Everyone has to start somewhere. As for my own lab at Rising Tide Brewing, it is a 7 × 9 ft. converted office and is not equipped with running water or a drain. While that lab may not be fancy, it is highly effective.

Basic equipment needed to start your lab:

- A cleanable worksurface for microbiology work; 3 × 2 ft. will do for a start
- A student microscope with 400–1000× magnification
- Incubator of some sort; this might be a DIY incubator made from a beverage cooler and light bulb
- Mini-refrigerator
- Glassware and accessories for microbiology work (see chapter 14)
- Stir plate
- Hydrometer and pocket stick-type pH meter
- Electric pressure cooker/sterilizer or autoclave

Our lab at Rising Tide started with that list above. We were able to get a lot done with that equipment coupled with good record keeping. As we grew and got better about consistently taking and documenting QCP measurements, we added equipment for what we consider important measurements, then added instruments for better precision and accuracy. For example, when we added our first brite tank we added a Zahm & Nagel CO_2 tester to measure carbonation. Several years later we upgraded to a canning line and added a combination DO and CO_2 testing instrument. This greatly improved our ability to consistently measure carbonation as well as allowing us to measure the performance of our new canning line.

Lab equipment we added over the first five years:

- Incubators for microbiology work
- Benchtop pH meter (easier calibration and better accuracy)
- DO meter; we obtained one when we moved from bottle conditioning to a bottling line
- Combination DO/CO_2 meter; we sold our previous DO meter to offset the cost of a new combination meter when we installed a canning line
- Digital densitometer
- Titration burette and stand
- Shaker table
- Autoclave

At Rising Tide, while we have added equipment to sharpen and expand our capabilities there has not been the opportunity in our expansions to build out a new lab. The result is that we still work in the original lab but have also spilled out onto the production floor. We moved equipment for what we call our "at-source QC" measurements on a lab table (stainless steel restaurant prep table) in full view of our brewery tasting room. This serves two purposes. Practically speaking, the location is convenient to our brewhouse, cellar, and packing hall for our production folks to take and record at-source measurements and make decisions based upon that data. From a marketing/company ethos standpoint, the table is public facing and reinforces to our guests, accounts, and distributing partners that we are a quality-focused brewery. The measures we use are displayed for them in full view, hiding nothing.

Ninety-five percent of the time, I can get the information I need in my own lab. When I come across an issue that I do not have the capability for, I reach out to my local university for help or my local craft brewery friends.[1]

[1] My thanks to Marcia and Lucy at the University of Southern Maine's QC2 Quality Control Collaboratory; and thanks to Zach, Jason, Mike, Heather, Hannah, Lee, and Karl at Allagash Brewing Company.

As for the sensory portion of Rising Tide's lab program, there is no dedicated space. We started out by taking over the tasting room bar before all staff meetings. We focused on shelf life, brand recognition, and flavor attribute spike recognition. Admittedly, those early days of sensory analysis were quite disorganized in terms of direction and record keeping. The process slowly evolved after we started to recognize and set standards around our brewery. We set up true-to-brand standards, introduced panel training sessions, and narrowed our flavor attribute spike training to fermentation-related attributes, all of this in conjunction with stepping up our data recording practices. While it was some work, I now have more trust in the data obtained from my small yet mighty sensory panel.

I share the story of our own lab to encourage you to start with what you have. Get creative and upgrade when you can.

ONE GOAL, DIFFERENT APPROACHES

The approach the we have taken with our lab at Rising Tide may not be the best approach for everyone. While researching for this book, I reached out to other folks in the industry to see how they have decided to set up their quality labs. While there were several different paths folks took, there were common threads throughout.

The Retrofit Approach

Fitting your lab into an existing space is a completely viable option to get you started. Having a door to separate you from the hustle and bustle of the brewery in order to do microbiological and other sensitive chemical measurements is a huge plus. I have heard of people retrofitting old offices, broom closets, even bathrooms. Whatever you need to do to get started, I say go for it! Once you know what you are measuring, you can prioritize what lives on the production floor on a lab table (or other workspace) and what stays in the lab. Think about the lab table or cart as a secondary lab space—it can also help keep quality at the forefront of the minds of other production employees. Training will go a long way when it comes to care and use of instrumentation. If your production employees are rough on equipment, your lab equipment may not last long. Train them not only on how it works, but also why it is important to the overall production of beer.

The Collector Approach

I have spoken to a few brewers that had limited space for their lab programs but had big visions for the future. While keeping up with the basic quality measurements (as discussed in earlier chapters of this book), they combed the internet and listings of lab wholesale equipment brokers for used equipment. When they found a good deal, they would scoop it up. When it was time to build out, they had a sense of what they wanted to add and the equipment to accomplish it. This influenced the build out as well as lowered the initial cost by utilizing used equipment acquired at a leisurely pace.

The Build Out Approach

Building out will likely be the best approach when your brewery is expanding or renovating a new production facility. When taking this approach, consult your business and build out plans while considering projected growth. While your quality system may not be complicated at the time of buildout, you will be growing, and you will likely be adding new lab equipment and staff.[2] Start by identifying the purpose and function of your quality lab:

- What equipment will be used?
- What analyses will be performed?
- Will the lab have a dedicated employee?
    - How many employees?
    - Will sensory analysis take place in part of the lab and how many people will the panel consist of?
- Does the lab require a workstation and computers?
- Will samples be stored or retained in the lab?

Design and Construction of Brewery Quality Labs from the Brewers Association is a great resource. You can follow this resource and work with your project manager/architect to come up with the best plan for your new lab. The ASBC's online "Grow Your Own Lab" resource also offers guidance on how to set up your lab or improve an existing one.[3]

It may not be cost prohibitive for a 10,000 bbl./year brewery to build out a lab of the same caliber as that of someone like Sierra Nevada Brewing Co. Keep these goals in mind when planning and leave some room to grow.

BUDGETING

There are many factors that influence budgeting for a lab, including equipment and consumables, plus any buildouts and new staff hires (who may possibly come with higher education qualifications). Budgets are a very personal thing, especially when planning a large buildout. If you have big plans for your brewery production goals, you should have big plans for your quality lab. A well furnished and organized lab will support your brewery in all facets for years to come. My point is, do not skimp on your lab program and its buildout.

As for equipment, start with the best you can afford. Another tip to filling out your lab on a budget is looking into university labs when they upgrade. You may be able to get a good deal on a piece of used equipment. Prioritize any planned upgrades, for example, any equipment where the measurement is difficult or cumbersome will probably get skipped, so these types of measurements get first consideration for instrument upgrades. Keep in mind it is easy to get "gadget fever" over a new piece of equipment or method, but be realistic in evaluating what process you are influencing by using it. Evaluate the actual purpose of the test, make sure it is critical to ensuring quality, and make sure there is someone in charge of running it before it ends up as a paperweight. As you acquire new production equipment, budget for any new instrument that will help confirm the equipment is working as you need. Examples include getting a DO meter for a new canning line; or, if you are anticipating a brewhouse change, like a larger system or adding a vessel, consider getting a spectrophotometer in advance to get baseline IBU and color measurements so you can start dialing in your new equipment as soon as possible.

Budgeting for personnel will depend greatly on your size. The Brewers Association again has a useful resource in the form of its biannual Brewery Operations Benchmarking Survey (BOBS). It provides a valuable set of tools specific to America's small and independent craft brewers, including average salaries by position. See also appendix H of this book, which contains a budget estimate in two parts. One is for breweries that offer beer for consumption only on the premises and an add-on part for breweries that package their beer for consumption outside of the brewery.

[2] The Brewers Association has a wonderful resource for folks who want to build out a new lab: "Design and Construction of Brewery Quality Labs," Brewers Association [website], accessed June 18, 2019, https://www.brewersassociation.org/educational-publications/design-and-construction-of-brewery-quality-labs/ [subscription required].

[3] "Grow Your Own Lab," American Society of Brewing Chemists (website), accessed December 7, 2019, https://www.asbcnet.org/lab/getstarted/Pages/growyourown.aspx.

LAB SAFETY

With your lab in place, it is important to think about safety just as you would in the brewhouse. The ASBC has a comprehensive lab safety check list.[4]

HAZARDS PRESENT IN BREWERY LABS

While not all hazards listed below may be found in all brewery labs, many will be.

1. **Chemicals** – toxins, flammables, caustics, corrosives, carcinogens, sensitizers
   a. Always wear appropriate personal protective equipment (PPE) when handling chemicals.
   b. Ensure all chemicals are clearly labeled and are accompanied with the proper safety data sheet (SDS).
   c. Do not store incompatible chemicals together.
2. **Biohazards** – antibiotics used to prepare microbiological media can be toxic even in small amounts
   a. Eating or storing food where these antibiotics are being used or stored should be strictly prohibited.
   b. Thoroughly clean all glassware after each preparation of media to prevent cross contamination of antibiotics.
   c. Prohibit the drinking of any liquid out of lab glassware.
      i. If preparing a VDK force test in flask, consider dedicating and permanently labeling that flask(s) for that purpose only.
3. **Pressurized gases** – CO_2, O_2
   a. Just like in the brewhouse, all gas cylinders should be secured in an upright position with a chain or clasp, away from any objects that could fall and strike them.
4. **Electricity** – damaged wiring, overloaded circuits, water
   a. Ground fault circuit interrupter (GFCI) outlets are required near water sources.
   b. Inspect power cords regularly.
   c. Know where the circuit breakers for the lab are in case of emergency.
5. **Glassware**
   a. Eye protection and heat-resistant gloves should be used if the glassware is heated or placed in any situation in which breakage may occur such as pressure or vacuum.
   b. Clean glassware after each use to prevent cross contamination
6. **Fire**
   a. Be aware of flammable chemicals, especially in the presence of ignition source such as a Bunsen burner or alcohol burning lamp.
   b. Have a fire extinguisher in the lab and know what kind of fires it is rated for.
7. **Other physical hazards**
   a. Know the physical hazards associated to each piece of lab equipment such as, autoclaves, fume hoods, centrifuges, all electric powered equipment, sharps, and needles.

[4] *ASBC Lab Safety Checklist*, American Society of Brewing Chemists [website], accessed December 5, 2019, https://www.asbcnet.org/lab/safety/Documents/ASBCLabSafetyChecklist.pdf.

22

INTRODUCTION TO ANALYSIS

You now have your lab set up. You put in an enormous amount of effort planning for quality, putting CCPs and QCPs in place, and you know what measurements you want to make. But when starting to set standards, you need to come up with some strategies to monitor and control those standards. The brewing, manufacturing, or even the scientific process can be described as the process of variable control. Every standard you will set has a variable or set of variables that affect it. The better control you have over these variables the better you can control your standards and final product. To do this, you should be able to measure variables either directly or indirectly.

> **Direct measurements** are made with a tool to directly ascertain the value of the parameter.
>
> - Example: measuring the temperature of fermentation using a thermometer or performing a yeast cell count in a sample using a microscope and hemocytometer.
>
> **Indirect measurements** occur when you take the measurement of one unit/object and use it to track another parameter.
>
> - Example: using pressure and temperature to determine the CO_2 volumes in a liquid.

VARIABLES

Variables are elements, features, or factors that are liable to vary or change. Any process input can be a variable. For example, the amount or temperature of your strike water and the temperature of the grain in the mash tun are variables that affect your mash temperature.

There are three common types of variable you will need to consider. The **independent variable** is the variable singled out by the experimenter that is changed in a controlled manner to test its effect on another variable. The effected variable is the **dependent variable**. A change in the independent variable directly causes a change in the dependent variable. The effect on the dependent variable is measured and recorded. If you were to graph your results conventionally, the independent variable is plotted on the *x*-axis and the dependent variable is plotted on the *y*-axis. The final type of variable are the **control variables**. A control variable is an element that is not changed throughout an experiment so that only the effects of the independent variable are observed and measured.

Let us walk through a simple thought experiment to demonstrate variables. Suppose you pull 3 liters of unhopped, pre-boil wort from your brew kettle. You separate the wort into three vessels and boil them all for the same amount of time. At the start of each boil, you add varying amounts of a single hop variety.

- Vessel 1 receives no hop addition
- Vessel 2 receives 2 g/L
- Vessel 3 receives 4 g/L

After cooling, you analyze each wort for IBUs (e.g., using *ASBC Methods* Wort-23a). Vessel 1 measures zero IBUs because there were no hops added. The measured IBUs in vessel 2 are greater than the measured IBUs in vessel 1 but less than vessel 3, which had the highest IBUs of the group (fig. 22.1).

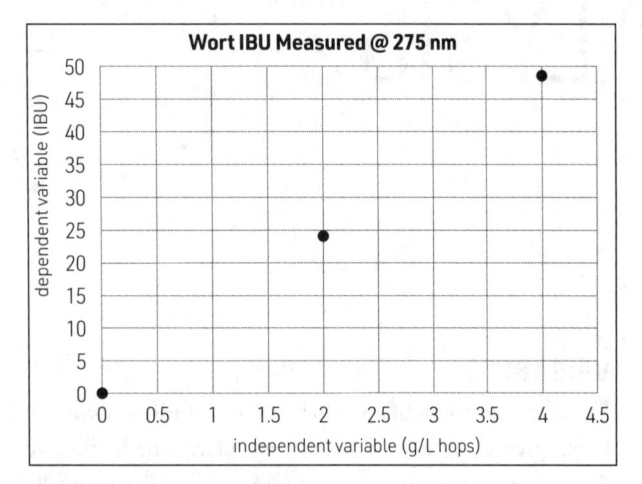

Figure 22.1. Independent and dependent variables plotted on a chart.

These predicted outcomes seem obvious, but the exercise was to show that

- the independent variable is the amount of hops added in g/L;
- the dependent variable is the measured IBU level; and
- the control variables are the initial conditions, volume, heating, and boiling time of the wort across the experiment, all of which should have been uniform between all three samples. While these variables are not included in the chart for analysis, they should be recorded to enable the experiment to be repeated if desired.

A point to make explicit is *always test one independent variable at a time.* If you change two independent variables at the same time, you will not be able to isolate which variable influenced the dependent variable (the outcome) or by how much. If you want to know the effect of changing multiple variables, you should conduct multiple tests where you focus on one independent variable at a time and compare the results to your control variable group.

PLAN-DO-STUDY-ACT CYCLE

With an understanding of variables, let us now look at strategies for controlling them. When you come across a parameter or process that is out of control or that you otherwise want to change, the plan-do-study-act (also known as the PDSA cycle, Deming cycle, or Shewhart cycle) is a valuable problem-solving model that uses four steps to implement a change. The cycle's simplicity allows it to be applied to just about any problem that arises or any change needing to be made. The PDSA cycle is cyclical in nature and can be applied to continuous improvement projects.[1]

1. **Plan:** Recognize an opportunity and plan a change.
2. **Do:** Test the change.
3. **Study:** Review the test, analyze the results, and identify what you have learned.
4. **Act:** Take action based on what you learned in the study step. If the change did not work, go through the cycle again with a different plan. If you were successful, incorporate what you learned from the test into wider changes. Use what you learned to plan new improvements, beginning the cycle again.

To understand this cycle a bit better, let us look at the example in figure 22.2. Here we see a control chart following mash pH over the last 25 turns. The recipe has been developed with water profile, grist bill, and calcium additions to yield a pH of 5.40, but recently the brewery changed malt suppliers. The normal distribution of data tells us that 99.87% of our data points should fall within 3 standard deviations on either side of the mean. Thanks to our control chart we can see the first 20 turns seemed to be reasonably in control. The last 5 turns alert us to a system out of control. This data initiates step one of the PDSA cycle. We can easily recognize the pH is skewing high and we now *plan* on a way to correct it.

There are no rules on how you *plan* the change, but it usually revolves a brainstorming session of some kind. It is obvious in this case that the new malt

[1] Nancy R. Tague, *The Quality Toolbox*, 2nd ed. (Milwaukee, WI: Quality Press, 2005), 390.

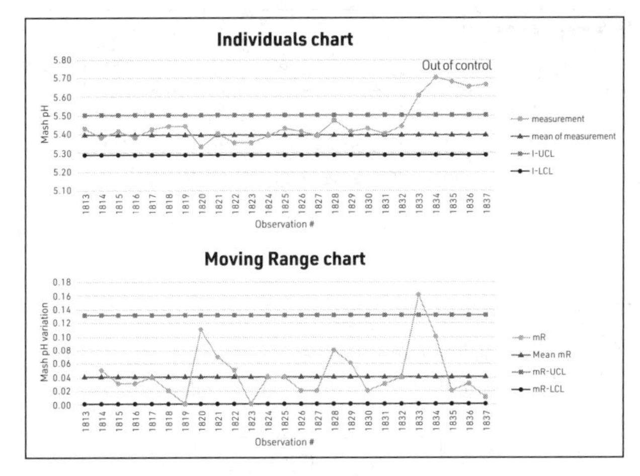

Figure 22.2. I-mR chart showing mash pH out of control.

supply is the likeliest cause of the elevated pH, but why? The primary goal at this stage is to identify as many causes of the defect as possible and then work on narrowing the root cause down. For this example, the ASBC has published a fishbone diagram that outlines different variables and their effect on pH, so this is a good place to start (fig. 22.3). Using the fishbone as a brainstorming aid, write down all likely variables that could influence the mash pH in that direction and surmise a root cause. Notice, for example, that gypsum additions will drive down the pH; maybe the gypsum addition is no longer sufficient to achieve your target mash pH of 5.40.

This will bring us to our next step in the PDSA cycle, *do*. Using the hypothesis from the *plan* stage, you decide to set up a benchtop experiment to test if the gypsum in the mash salts needs to be adjusted to achieve your target mash pH.

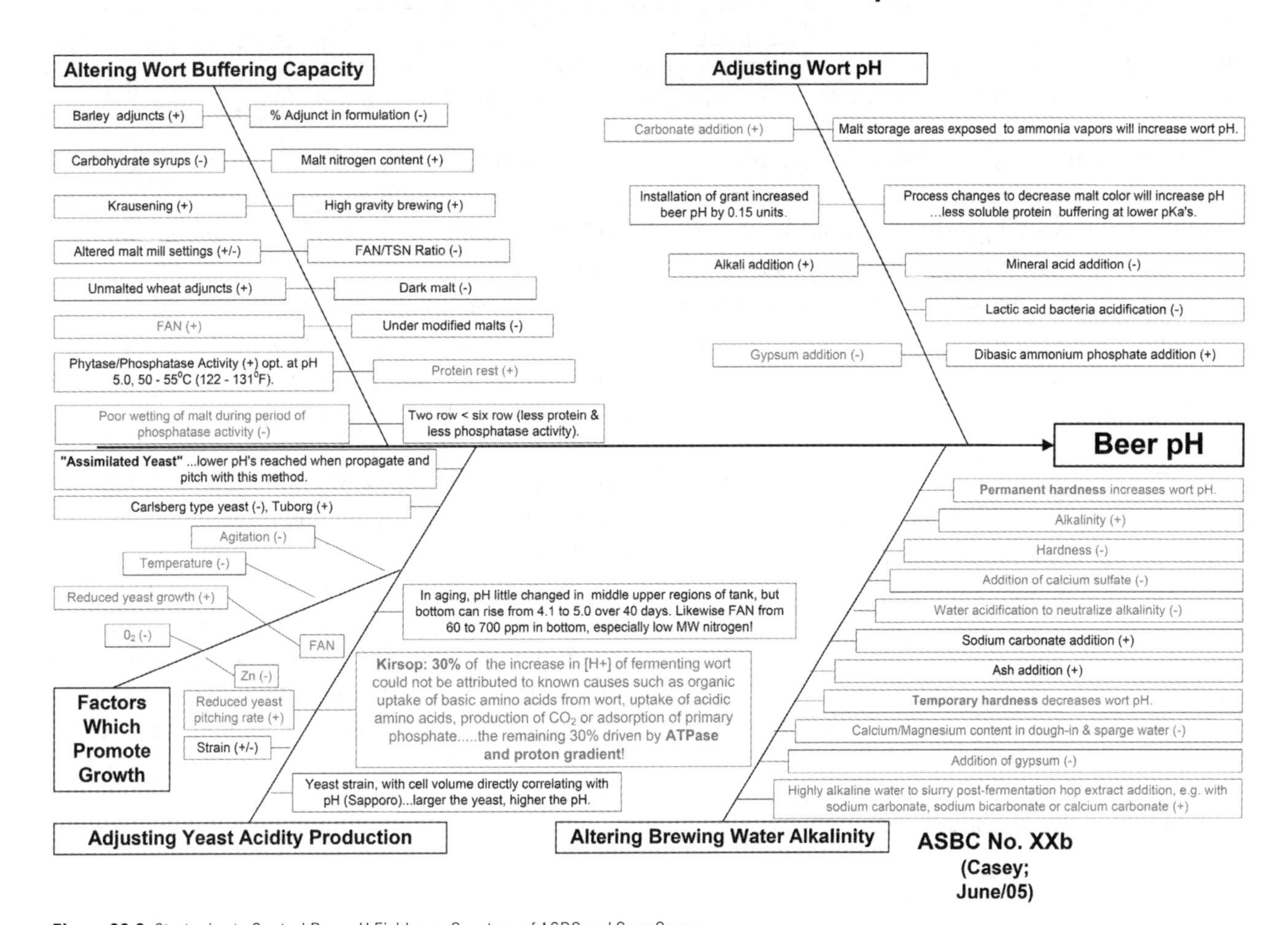

Figure 22.3. Strategies to Control Beer pH Fishbone. Courtesy of ASBC and Greg Casey.

Elevated Mash pH PDSA Experiments

Merritt Waldron | Rising Tide Brewing Co | 2/17/2019

Background and objective: A PDSA cycle was initiated after a deviation in mash pH was noticed in our control charts. The brewers and lab staff used the ASBC Fishbone diagram XXb Strategies to Control Beer pH as a guide in an effort to determine the root cause. Since our malt supplier has changed it is suspected that our gypsum addition needs to be adjusted to achieve our target pH.

Independent variable: Gypsum addition

Dependent variable: Mash pH

Control variables: Grist bill recipe, strike water ratio, and strike temperature

General procedure: Make six identical benchtop mini-mashes (100 g each) ensuring control variables are the same for each. Split into three groups, one for each independent variable tested. Group 1 is original recipe formulation. Group 2 gets a 50% increase in gypsum. Group 3 gets a 100% increase in gypsum.

Group 1 recipe	Group 2 recipe	Group 3 recipe
• 100 g grist • 0.20 g $CaSO_4$ • 270 mL strike water • 168°F strike water temperature	• 100 g grist • 0.30 g $CaSO_4$ • 270 mL strike water • 168°F strike water temperature	• 100 g grist • 0.40 g $CaSO_4$ • 270 mL strike water • 168°F strike water temperature

Results

Group 1		Group 2		Group 3	
Trial	Mash pH @68°F (20°C)	Trial	Mash pH @68°F (20°C)	Trial	Mash pH @68°F (20°C)
1	5.66	1	5.58	1	5.44
2	5.65	2	5.56	2	5.42

Observations: The mash pH seemed to be closer to the target mash pH when increasing the amount of gypsum for the mini-mashes. We will trial a 100% increase of our current gypsum dosing rate for the next mash.

Onto the next step, *study*. Based on your observations from the elevated mash pH experiments (see sidebar) the increased gypsum additions seemed to bring the pH closer to the target range. Based on your findings, the next action you take is to double the amount of gypsum added to the full-scale mash on the next brew. Using what you find from the *study*, you will enter the final stage and *act* by initiating a full-scale recipe change to the brew and monitor the changes closely.

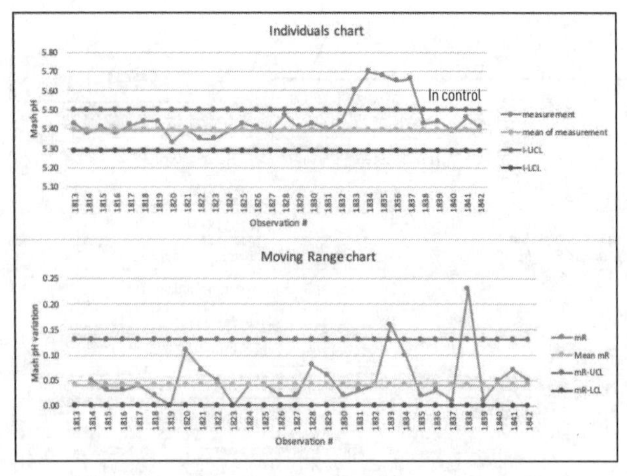

Figure 22.4. Mash pH back in control after salt adjustments.

Monitoring over the next five turns it looks as though the mash pH has been corrected (fig. 22.4). If this was not the case, I would go back to the planning stage and start the PDSA cycle again.

Brewing is a business of controlling variables. By working with independent, dependent, and control variables, you can use strategic problem-solving methods such as the PDSA cycle to correct and find issues with systems or processes that seem out of control. In the next two chapters we will go in-depth on working with variables and look at how to build a control chart identical the ones we just saw in figures 22.2 and 22.4 to visualize processes around the brewery.

23

DATA COLLECTION AND DATA QUALITY

In a busy, craft brewery it will be difficult for a single person working in the quality lab to take, record, and analyze every process measurement on their own. Imagine taking all the brewhouse measurements while simultaneously checking DO on the packing line and pulling daily gravity and pH measurements. While you may be able to keep up for a day or two, I guarantee that you will not get to analyze that data in a meaningful way. By assigning these duties to the process operator (brewer, cellar person, packaging crew), we can introduce the idea of "at-source quality." At-source quality is the act of assigning in-process quality measurements and data recording to the individual performing the task at hand. This frees the quality lab staff to accomplish other specialized tasks such as microbiology, in-depth chemical analysis, sensory analysis, and, most importantly, data analysis. It gives the process operator the information needed to positively influence the process and keeps the operator invested in quality. This also removes the lab person from being the proverbial messenger of bad tidings, which helps quell unneeded tension between operators and lab staff.

There are some obstacles that arise from having multiple operators making at-source quality measurements. A significant obstacle is variation in measurement or process by multiple operators. This is caused by a lack of documented SOPs, or inadequate training and supervision (think back to your GMP manual). Refer back to part one of this book to help address these issues.

Part of the standardization process is agreeing on the exact parameter to be observed and ensuring that everyone is measuring the same thing. Next, decide both how often the events will be observed (the frequency) and over what total period (the duration) where applicable.[1] The important thing to note is to measure the parameter the same way every time. This will lead to useable, consistent, and relatable data, regardless of accuracy and precision (more on this in a second). So, how can you tell if the data you are collecting is any good?

ACCURACY AND PRECISION

You put a lot of effort into standardizing measurements so that you and your staff can record accurate and precise data. While accurate and precise sound like the same concept they are two slightly different things. **Accuracy** refers to how close the measurement is to the "true" or accepted value. **Precision** refers to the repeatability of the measurement.

1 "Quality Priority Pyramid," Brewers Association [website], accessed July 18, 2018, https://www.brewersassociation.org/educational-publications/quality-priority-pyramid/.

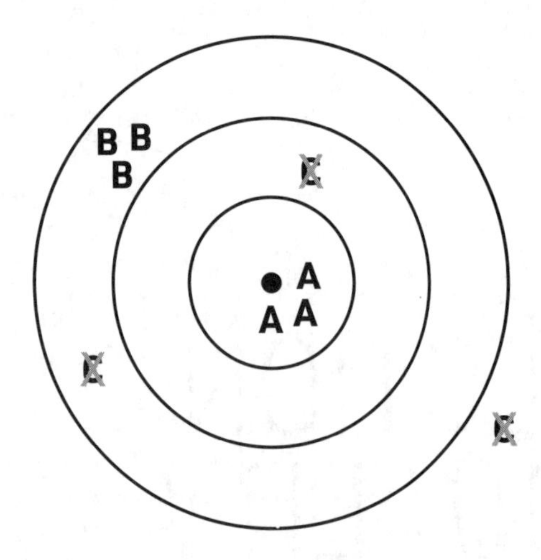

Figure 23.1. Accuracy and precision over a target.

In figure 23.1, A is a grouping of three measurements that are both accurate and precise showing that the measurement is close to its true value and repeatable. Notice that not every value for A is the same, which is an example of normal variation.

Grouping B in figure 23.1 is an example of a group of measurements that show precision but not accuracy. This is a good sign that the measurement is repeatable but it needs slight adjustment or calibration. If a calibration cannot be performed, this may still produce useable data if the bias is consistent. Think of a slightly heavy hydrometer, for example—it will always read low but it will be consistently low so data can be compared provided the values are taken with that same hydrometer. The absence of accuracy will be a problem when you get a new hydrometer or when comparing values to those taken using different instrumentation or by different labs.

The measurements in group C are neither accurate or precise and are therefore not useable. If this happens, the measurement method should be evaluated and the sources of error tracked down before making decisions based on that measurement.

SOURCES OF ERROR

Error is inherent in the measurement process, even if your measurements are accurate and precise. Notice that grouping A from figure 23.1 is accurate and precise, but each measurement is not stacked on top of each other on the bullseye. That would be like three holes in one on the same par-four hole, a highly improbable event for anyone (especially for me). But in grouping A all the measurements are on the green and in good shape, so to speak.

Sources of error in any measurement process can be described in the following way:[2]

Total error = sampling error
+ systematic error (bias, reproducibility)
+ repeatability error (replicative)

As you can see, reducing any one of these sources of error will reduce the total error in measurement. Understanding where these different types error come from will help you avoid them and keep your processes in control.

Sampling errors often occur because you are taking a small sample from a large batch (see "Avoiding Sampling Errors" as an example of what I mean). A strategy to reduce sampling error is to sample from multiple points, crosschecking the results and seeing if they agree. Contamination is also a common sampling error. Reduce this by "rinsing in" samples to your vessel, being sure that the all the previous sample and cleaning water is removed (see "Contamination Mitigation").

AVOIDING SAMPLING ERRORS: A CASE STUDY

One experience we had at Rising Tide illustrates how easy it can be to introduce a sampling error. When taking a final cell count before a beer was to be transferred to brite (a QCP of ours), we did not have a standardized protocol for drawing that sample. Because of the geometry of our fermentors, some brewers were obtaining higher than expected cell counts while others were finding themselves on target. As it turned out, those hitting the target were blowing out all the yeast that built up behind the sample valve. To resolve this issue we instituted a standardized sampling protocol (an SOP) for this measurement that included blowing out the yeast buildup prior to taking the sample, which reduced the error in measurement of cells counted pre-transfer. This brought all brewers closer to the true value of yeast in suspension.

2 Michael Lewis and Tom Young, *Brewing*, 2nd ed. (New York: Kluwer Academic/Plenum Publishers, 2002), 117.

CONTAMINATION MITIGATION

Contamination in this case refers to the liquid sample being altered by the remnants of the previous sample. For example, you have a beaker that was just washed out with water between gravity samples and you did not dry it. You immediately fill the beaker with 100 mL of sample. Because the beaker was not dried there is residual water clinging to the walls of the vessel, let us say 2 mL. That 2 mL of water will dilute your 100 mL sample by just under two percent. This is not an insignificant result. Think of chugging a glass of water and putting it down on the table to let it stand for a minute, you will see water collect at the bottom.

Rinsing in is a technique used to mitigate the influence of sample contamination when sampling many liquids in quick succession. A small amount of "sacrificial" sample is allowed to flow into the vessel and then immediately discarded. In that same vessel the sample to be analyzed is then collected and measured. This "tempers" the vessel by reducing the influence of the previous contents. The act of rinsing in can be likened to the act of preheating the mash tun.

Systematic errors result in values being off target. Compare grouping A to grouping B in the earlier example (fig. 23.1). This kind of error can be reduced by proper calibration of the instrument. Systematic error can also be mitigated by specifying the instrument to be used for measurement. In the case of the "heavy" hydrometer mentioned in the previous section, make sure it is the only one used across all density measurements in the brewery and correct for the known bias. Better yet, of course, is to replace it with an accurate hydrometer. Systematic error is not detected by statistical analysis, but it can be detected by measuring a known standard of precise concentration or value.

Repeatability errors are those that cannot be explained by sampling error or systematic error. They are attributed to random variation. The lower this error, the lower the variation in the results obtained by repeated measurement of the same sample.[3] This type of error can be detected by statistical analysis.

[3] Robert Niles [website], "Standard Deviation," accessed June 15, 2019, https://www.robertniles.com/stats/stdev.shtml.

MEAN, NORMAL DISTRIBUTION, AND STANDARD DEVIATION

With all this talk of error, how do you know what the true value of your measurement is? How can you be sure? To determine the true value of a measurement, we can employ some statistical methods to get the best estimate of the true value. This will alert you to random error and instill confidence in your measurements. We need to get a bit mathematical here, but I will do my best to make it short, sweet, and worth your while.

It is common practice in a lab to take multiple measurements of the same sample. This is good practice in the brewery as well because one data point will not tell you much. By taking multiple measurements, we can use the *mean*, or average (represented by $\overline{x}$), to give the best representation of the "true" value over a series of measurements. We have been figuring out averages since grade school:

$$x_{\text{best}} = \overline{x} = (x_1 + x_2 + \cdots\ x_n)/n$$

where n is the total number of measurements.

For example, having identified all sources of systematic error, suppose you use your density meter to check the density of the knockout wort and obtain its starting gravity. You take five density measurements of your knockout wort sample, $x_1, x_2, \cdots x_5$, therefore, n is 5.

12.0, 12.4, 12.2, 12.4, 12.0 (in degrees Plato, °P)

You can find the mean of these five values to get your best estimate ($\overline{x}$) for the density of the wort:

$$\begin{aligned}\overline{x} &= (x_1 + x_2 + \cdots\ x_n)/n \\ &= (12.0 + 12.4 + 12.2 + 12.4 + 12.0)/5 \\ &= 12.2°\text{P}\end{aligned}$$

You notice variation among the measurements in your sample set above. This is not insignificant and can be described by the *standard deviation* (SD), which expresses how far each measurement deviates from the mean. To determine the SD, start by subtracting each measurement value (x_i, where i is an individual measurement, x_1, x_2, etc.) from the mean ($\overline{x}$) and obtain the deviation (d_i, where i corresponds to the trial number in x_i) and average deviation ($\overline{d}$).

$$d_i = x_i - \bar{x}$$

Using the data from your starting gravity data set, where $\bar{x}$ =12.2, you can calculate d_i:

Trial number i	Measured value x_i	Deviation $d_i = x_i - \bar{x}$
1	12.0	-0.2
2	12.4	0.2
3	12.2	0.0
4	12.4	0.2
5	12.0	-0.2
	sum of x_i = 61.0	sum of d_i = 0.0
	$\bar{x}$ = 12.2	$\bar{d}$ = 0.0

With a result of zero for deviation, you can see that the average deviation is not a helpful way to determine the reliability of your measurements.[4] This looks like a dead end but it is not, it is just a consequence of this data set, which I chose carefully to exhibit the difference between *average deviation* ($\bar{d}$) and *standard deviation*, which is usually represented by sigma (σ). To resolve this issue of a zero sum and continue to the standard deviation you need to create a set of positive numbers. This is accomplished by squaring all the deviations (d_i^2) and then taking the average of those values (this value is the *variance*).

Trial number i	Measured value x_i	Deviation $d_i = x_i - \bar{x}$	Deviation squared (variance) d_i^2
1	12.0	-0.2	0.04
2	12.4	0.2	0.04
3	12.2	0.0	0.00
4	12.4	0.2	0.04
5	12.0	-0.2	0.04
	sum of x_i = 61.0	sum of d_i = 0.0	sum of d_i^2 = 0.16
	$\bar{x}$ = 12.2	$\bar{d}$ = 0.0	$\bar{d}_i^2$ = 0.03

If we then take the square root of that result ($\sqrt{\bar{d}_i^2}$) we get the standard deviation (σ_x) of our data set for *x*.

[4] John R. Taylor, An Introduction to Error Analysis the Study of Uncertainties in Physical Measurements, 2nd ed. (Sausalito, CA: University Science Books, 1981), 99.

$$\sigma_x = \sqrt{\bar{d}_i^2} = \sqrt{0.03} = 0.17$$

Therefore, the standard deviation, σ_x, for our starting gravity measurement equals 0.17.

The definition of the standard deviation is commonly written as:

$$\sigma_x = \sqrt{\frac{1}{N}\sum_{i-1}^{N} (d_i)^2}$$

Do not let this notation intimidate you. It just represents all of the steps we completed in this section in a neat little package.

Standard deviation (σ) tells you how widely the values in a set are spread apart. A large value for σ tells you that the measurements are more variable. A small σ tells you the measurements are tightly bunched together around the mean, such as grouping B in figure 23.1. This is your "ah-ha" moment, because if you know you have eliminated all systematic error, in other words your instrument is reading accurately, then a tight standard deviation tells you that your precision is on point. Putting these facts together, you can be confident that the actual value of your starting gravity is 12.2°P.

NORMAL DISTRIBUTION OF DATA

The next obvious question is, how do you know if your standard deviation is indeed tight? In statistics, a **normal distribution** refers to the distribution of data around the mean. In a data set with a normal distribution, most of the data points will be close to the mean, while a relatively small number of data points tend to one extreme or the other. When data are plotted so that the *x*-axis represents the value of the measurement and the *y*-axis is the frequency with which the measurement occurs, a normal distribution produces a *bell curve*. The vertex of the curve (highest point on top of the bell) sits above the measurement's mean value on the *x*-axis (fig. 23.2). The area under the curve shows that your data points fall most frequently around or close to the mean, and less frequently as you move further away from the mean.

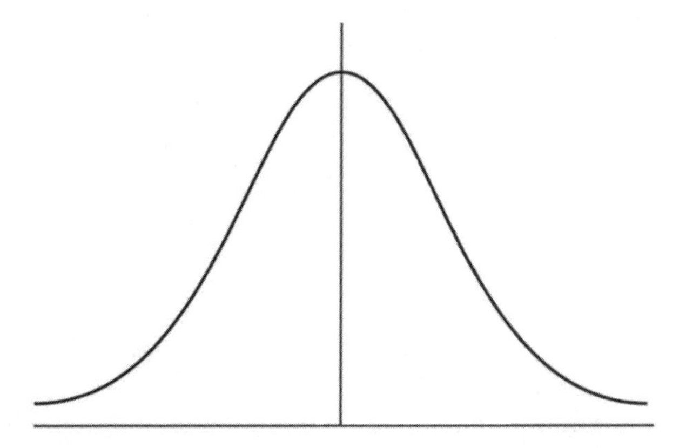

Figure 23.2. Normal distribution about the mean. The area under the curve shows that your data points fall most frequently around or close to the mean (represented by the line that is drawn down from the vertex of the curve), and less frequently as you move further away from the mean.

As stated above, the standard deviation (σ) tells you how widely the values in a data set are spread. There is an empirical rule for a data set that follows a normal distribution, which is that nearly all data will fall within three standard deviations of the mean. The empirical rule can be broken down into three parts, also known as the three-sigma rule:[5]

- 68% of the data falls within one standard deviation from the mean;
- 95.4% of the data falls within two standard deviations of the mean;
- 99.7% of the data falls within three standard deviations of the mean.

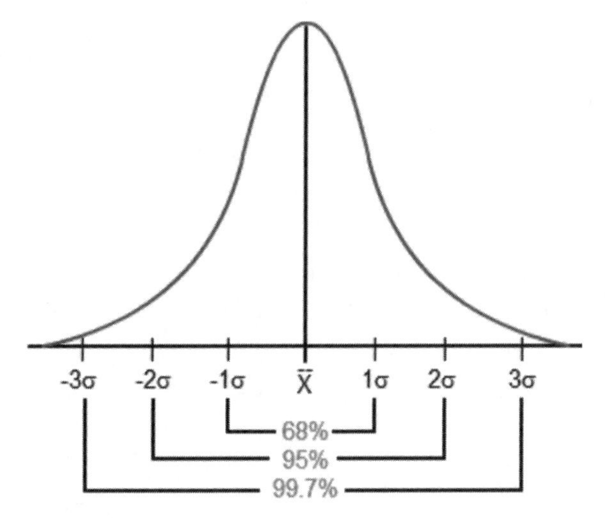

Figure 23.3. Curve of a normal distribution from a sample data set with sigma identifiers indicated.

You can visually see this rule represented under the standard normal curve (fig. 23.3). Not all data sets with a normal distribution will have graphs that look this perfect—some will have relatively flat curves, others will be steep, sometimes the mean will lean a little bit to one side or the other. But all data that follows a normal distribution will have something like this same bell curve shape.[6] For example, when the data values are spread apart and the bell curve is relatively flat, that tells you have a relatively large standard deviation and, therefore, your data is less precise. When the values are tight together the bell curve is relatively steep, meaning you have a smaller standard deviation and more precise data. If your data are not fitting a bell curve shape there is likely a variable out of control. Apply root-cause analysis techniques, and investigate the measurement and process variables using skills we learned in chapter 22.

Understanding standard deviation and its relation to the normal distribution opens up a world of possibilities for your data analysis. It allows you to validate measurement processes, determine uncertainties, and gives you the ability to start control charting. We are now dipping our toes in the world of statistical process control.

5 Taylor, *Introduction to Error Analysis*, 137.

6 Robert Niles, https://www.robertniles.com/stats/stdev.shtml.

24

DATA ORGANIZATION AND ANALYSIS

To get into the world of statistical process control you are going to need to collect some data. Luckily, breweries produce loads of data every day. No matter how you decide to collect that data, keeping it organized and quickly retrievable is the first step to being able to analyze it in a meaningful way.

Brewing is a batch process and lends itself nicely to keeping batch records, often referred to as brew logs. These process records aid the standardization of data collection in several ways. First and foremost, they serve as a log for at-source measurements and observations. They help ensure uniform and accurate information is recorded and they act as a visual checklist accompanying the SOP to help keep the operator on track. Each brewery develops its own style of brew log. Many craft breweries use paper brew logs, which is perfectly fine, but I would encourage you to go digital as soon as you can. Digital logs are easy to access from anywhere in the brewery or at home, and the added advantage is you can cut and paste data to help with analysis.

DEVELOPING BREW LOGS

There are some basic things you should consider when developing brew logs for your process:

1. Include batch process identifiers, such as a batch or date code of some kind.
2. Include data points directly from your sampling plan, especially at-source quality measurements.
   a. Ensure there is ample room to record data legibly.
3. Display data in a logical order from the beginning of the process to the end.
4. If QCP targets are established (e.g., original extract), include them on the brew sheet near where the process measurement is to be recorded so the information is readily available to all.
5. Will you include lot numbers for traceability?
   a. Again, ensure there is ample room to record data legibly.

Including these basic standards, your brew log should reflect your sampling plan and follow each batch from ingredients to package (see Brew Log Example appendix J). Most recipe development software will provide a brew sheet, which is helpful for brew day but you will likely have to develop a sheet or two to follow your beer through the cellar and beyond. A spreadsheet or word processing program of your choice will help develop process sheets beyond brew day. Information from your sheets should be entered into a database to be analyzed. There are many data analysis tools out there.

ANALYZE YOUR DATA

With your sampling plan and brew logs in place you can start to organize your data, putting it to work for you by using a statistical approach to variation management. Your goal is to visualize the data you have collected and determine something meaningful about it. Is the process behaving as expected? Is the process in control? Are your measurements accurate? Love them or hate them, a simple spreadsheet program will work just fine to get you started on this journey. While there are many software tools with QA/QC functions (see sidebar) to aid you in collection and data analysis, they often come with a substantial subscription cost. Spreadsheet programs like Microsoft Excel® are relatively inexpensive and universally available, contain many built-in functions, are relatively easy to use, and are familiar to most people. We will focus on them for this text.

BREWERY QC SOFTWARE EXAMPLES

OrchestratedBEER ("OBeer")
https://www.orchestratedbeer.com/

VicinityBrew
https://www.vicinitybrew.com/

Crafted ERP
https://www.craftederp.com/

Brew-Q
http://kgdatasolutions.com/

Getting Data into a Spreadsheet

No matter how you decide to collect and organize your data, in order for you to visualize it, it will eventually have to find its way to a spreadsheet. At Rising Tide Brewing we would collect all batch data on brew sheets like the ones you see in appendix J. That means I had to schedule a few hours each week to enter data into spreadsheets, but the time spent was totally worth it. Using that data, the brewery was able to reduce costs and improve efficiency in several processes. Most importantly, we caught deviations and corrected them before it was too late and beer was lost. Some instrumentation has the capability to digitally record data as the measurement is taken and that can be imported into your spreadsheet program. Rising Tide is moving more processes to user-entered digital data logging in order to reduce the time taken for manual data entry. Start digital if you can, but, if not, just start somewhere.

Simple rules
Group like data
Tables are your friend

There are many ways to organize your data on a spreadsheet and make it look great, but when entering data to make charts and graphs stick to the basics. Assign each measurement parameter to a column and always label your data. Let us look at an example of fermentation tracking (fig. 24.1). You can see all measured parameters, starting with time and ending with pH, have their own column. It is nothing fancy and totally attainable, just data ready to be visualized.

	A	B	C	D	E	F	G	H
1	Hour	Date	Time	Temp (F)	Set Temp (F)	SG	PH	Brewer
2	0	2/13/19	8:25	64	65	1.0393	5.35	BA
3	24	2/14/19	8:41	64.6	65	1.0346	4.69	AC
4	48	2/15/19	8:55	64.8	65	1.0220	4.55	MH
5	72	2/16/19	9:05	66.7	68	1.0124	4.45	MH
6	96	2/17/19	9:00	67.7	68	1.0070	4.51	MH
7	120	2/18/19	8:35	67.2	68	1.0071	4.55	ML
8	144	2/19/19	8:30	66.9	68	1.0071	4.5	ML
9	168	2/20/19	8:55	67.2	68	1.0068	4.51	AC
10	192	2/21/19	9:20	67.4	68	1.0064	4.54	MH
11	216	2/22/19	8:47	67.6	68	1.0064	4.5	AR
12	240	2/23/19	9:49	61.9	55	1.0066	4.58	ML
13	264	2/24/19	8:26	55.2	65	1.0067	4.62	ML
14	288	2/25/19	8:05	58.1	65	1.0064	4.64	MO
15	312	2/26/19	7:50	58.8	65	1.0065	4.66	BA
16	336	2/27/19	8:19	58.6	65	1.0064	4.63	ML
17	360	2/28/19	7:59	39.4	32	1.0065	4.69	ML
18	384	3/1/19	8:24	37.3	33	1.0061	4.7	AC
19								
20								

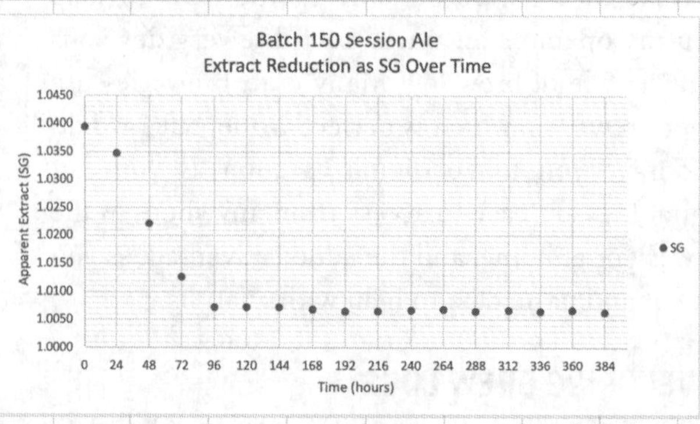

Batch 150 Session Ale
Extract Reduction as SG Over Time
Apparent Extract (SG)
Time (hours)
SG

Figure 24.1. Example of fermentation data organization and simple fermentation chart.

Charting Best Practices

For some of us, it has been quite a while since we have made a chart or plot. A chart is a general term for a graphical representation of data. They are used to understand relationships between data points and should be easier to read than the raw data itself. There are many kinds of charts and no single chart is best for every kind of data. We will go over only two examples, but you will notice similarities between them. For your data to be easily understood by yourself and others, it must contain the following basic elements (fig. 24.2).

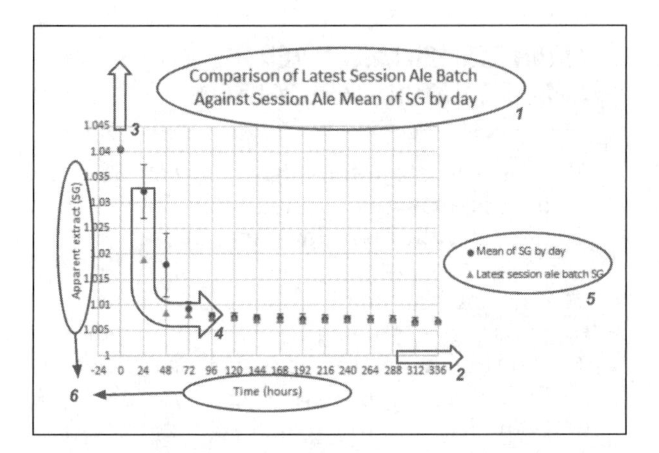

Figure 24.2. Example showing the basic elements of a well-labeled chart.

Title (1): The title offers a short explanation of what is in your chart. This helps the reader identify what they are about to look at. Explain the variable being compared, even something as simple as *y* (dependent variable) versus *x* (independent variable) will tell the reader a lot.

x-axis (2): The *x*-axis will always be arranged horizontally in a two-dimensional chart like the ones we will work on. The values along the *x*-axis will represent the independent variable, such as time or a batch identifier.

y-axis (3): The *y*-axis will always be arranged vertically in a two-dimensional chart like the ones we will work on. The values along the *y*-axis will represent the dependent variable.

Data (4): The displayed data is the meat of your chart and shows the data that has been collected. Data points are typically represented as a symbol such as a dot, tick mark, line, or a bar.

Legend (5): The legend tells you what the data points or bars represent. Just like on a map, the legend helps the reader understand what they are looking at.

Axis labels and units (6): Axis labels and units are essential to the readability of your chart. Each axis label (*x* and *y*) describes the variable being charted. All too often the units are forgotten. Units are important to the reader to describe the scale of the information provided. For example, consider using the *x*-axis label of time: when examining the data point at 192, does time refer to seconds, hours, days, fortnights, or years? You can see the number has little meaning without specific units.

Charts Example: Process Over Time

There are many processes that we monitor over time, and probably the most obvious and important is fermentation. You can tell a lot about your yeast and how it performs by visualizing fermentation data over time. Using the data from figure 24.1 (see above) we can plot the daily apparent extract as a drop in specific gravity over time, with time in hours on the *x*-axis and specific gravity on the *y*-axis. This plot is shown in figure 24.3.

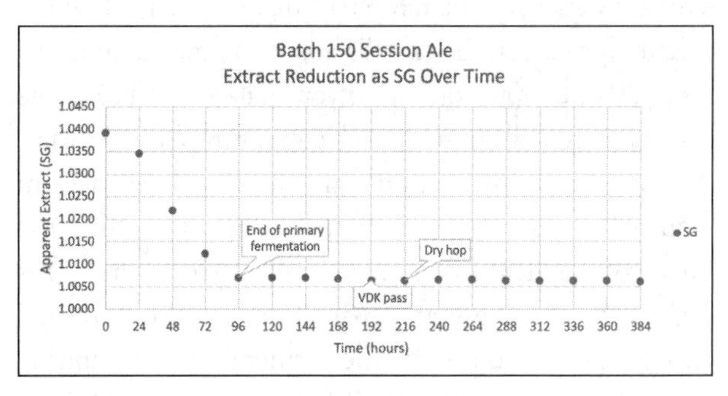

Figure 24.3. Chart of apparent extract reduction over time.

We can clearly see how our yeast performed from the curve in figure 24.3. You can easily do this for any process that you are interested in that is measured at regular intervals over time.

Now that you know how to visualize one batch at a time, let us take it another step further. Suppose your session ale has become popular and you are brewing it many times a year. You may want to look at how the fermentation of your latest batch stacks up to your previous fermentations of that brand. Assuming you

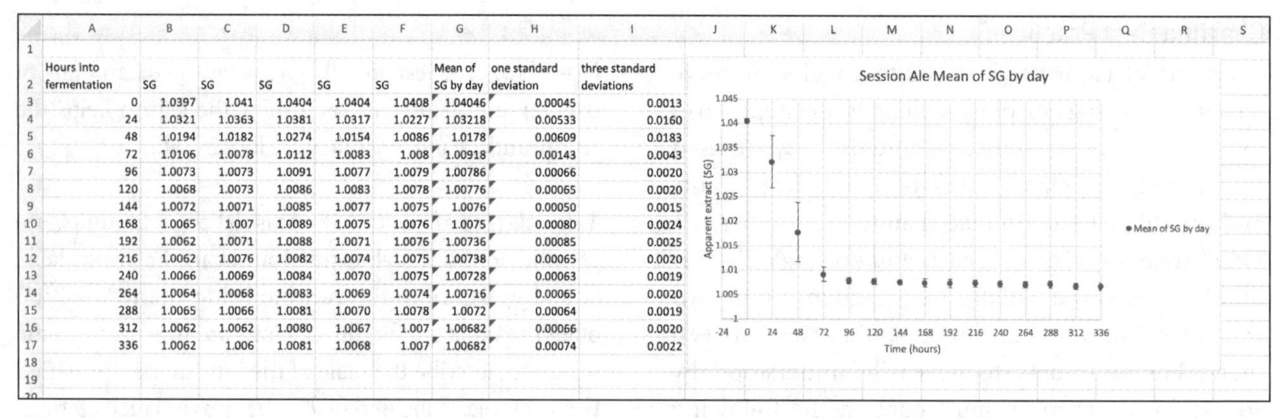

Hours into fermentation	SG	SG	SG	SG	SG	Mean of SG by day	one standard deviation	three standard deviations
0	1.0397	1.041	1.0404	1.0404	1.0408	1.04046	0.00045	0.0013
24	1.0321	1.0363	1.0381	1.0317	1.0227	1.03218	0.00533	0.0160
48	1.0194	1.0182	1.0274	1.0154	1.0086	1.0178	0.00609	0.0183
72	1.0106	1.0078	1.0112	1.0083	1.008	1.00918	0.00143	0.0043
96	1.0073	1.0073	1.0091	1.0077	1.0079	1.00786	0.00066	0.0020
120	1.0068	1.0073	1.0086	1.0083	1.0078	1.00776	0.00065	0.0020
144	1.0072	1.0071	1.0085	1.0077	1.0075	1.0076	0.00050	0.0015
168	1.0065	1.007	1.0089	1.0075	1.0076	1.0075	0.00080	0.0024
192	1.0062	1.0071	1.0088	1.0071	1.0076	1.00736	0.00085	0.0025
216	1.0062	1.0076	1.0082	1.0074	1.0075	1.00738	0.00065	0.0020
240	1.0066	1.0069	1.0084	1.0070	1.0075	1.00728	0.00063	0.0019
264	1.0064	1.0068	1.0083	1.0069	1.0074	1.00716	0.00065	0.0020
288	1.0065	1.0066	1.0081	1.0070	1.0078	1.0072	0.00064	0.0019
312	1.0062	1.0062	1.0080	1.0067	1.007	1.00682	0.00066	0.0020
336	1.0062	1.006	1.0081	1.0068	1.007	1.00682	0.00074	0.0022

Figure 24.4. Mean value of SG by day and standard deviation. Error bars in the chart represent one standard deviation from the mean.

are attempting to control all your major input variables (mash variables, boil variable, starting gravity, yeast cell count, etc.) you can use historical data collected from previous batches to build a more informative chart, which, in this example, we will title "Mean of SG by day."

To build the mean of SG by day chart, start by calculating the mean of your last five session ale fermentations by day (you can add as many fermentations as you like, the more the merrier). In other words, calculate the average of your starting gravity measurements, then the average of your 24-hr. gravity measurements, and so on.[1] Since we just learned about standard deviation and the normal distribution (covered in chapter 23) we can use that skill to create a visual aid in a new chart that helps us determine where our current batch falls about the mean. To do so just add one more column to your data set and calculate the standard deviation for each day's average (fig. 24.4).[2]

Using these instructions and the calculated data in figure 24.4 you can create a new chart by plotting the mean of the measured gravity by day against time. This represents the best approximation for the expected gravity at that point in the fermentation. Assuming your data has a normal distribution, the likelihood of the latest batch's specific gravity falling within one standard deviation is 68%. I have chosen one standard deviation as a marker here not because of a statistical significance but for the inferences I can make about the data. Knowing that 68% of your data points should fall within one standard deviation of the mean, this visual aid (including the error bars that represent 1 SD) can quickly highlight a potential anomaly that may be worth investigating. With the Mean of SG by Day chart you can compare subsequent batches of the same brand simply by overlaying the specific gravity values of your most recent batch as a new series. This allows you to compare it to the best approximation of the true value (mean) for that brand.

Looking at the overlay chart (fig. 24.5) you can quickly see the latest batch is well outside one standard deviation. The specific gravity at 24 hours is more than 2 standard deviations from the mean, which should raise a red flag as a possible anomaly. Using this chart and referring to the batch data you can start to try and determine what went wrong. Think about

MEAN FERMENTATION VERSUS SPECIFIC BATCH STEP-BY-STEP

1. Organize data in spreadsheet so calculations are easily conducted and clear.
2. Calculate the mean (average) of each day's gravity reading.
3. Calculate the standard deviation of each day's gravity reading.
4. Insert scatter chart and plot Mean of SG By Day versus Hours Into Fermentation Columns.
5. Insert custom error bars for each days data point using the standard deviation column for the value of each day. This is our visual aid.
6. Overlay data for the most recent batch by plotting it as a new series.
7. Perform visual comparison.

1 If using Google Sheets™ or Microsoft Excel®, this is easily accomplished by using the **AVERAGE** function. Both apps use the syntax =AVERAGE(value1, [value2, ...]) where [value2, ...] is optional.

2 In the latest version of Microsoft Excel, this can be done using the **STDEV.P** function. Google Sheets and earlier versions of Excel prior to 2010 use **STDEVP**. The syntax for both functions is the same, =STDEVP(value1, [value2, ...]) or =STDEV.P(value1, [value2, ...]).

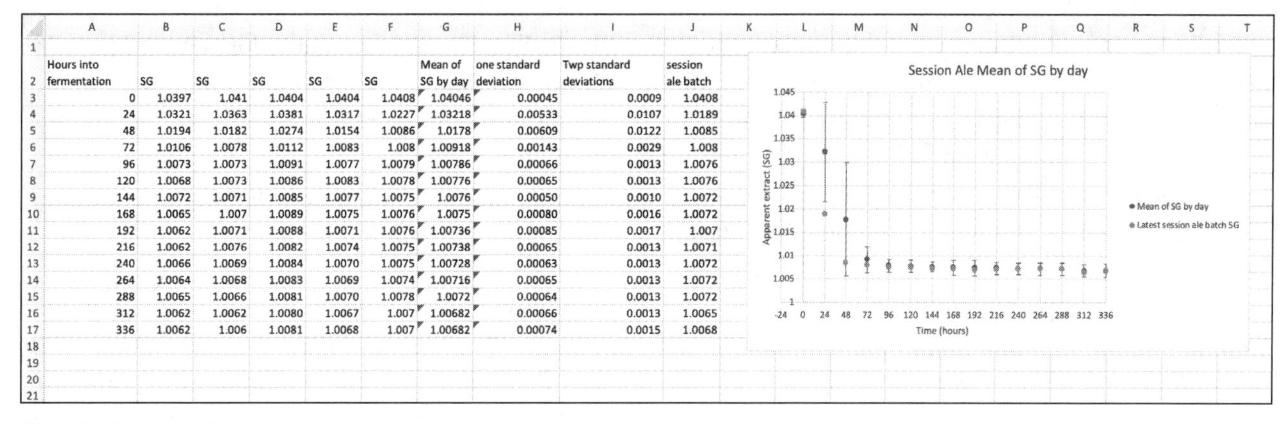

Hours into fermentation	SG	SG	SG	SG	SG	Mean of SG by day	one standard deviation	Twp standard deviations	session ale batch
0	1.0397	1.041	1.0404	1.0404	1.0408	1.04046	0.00045	0.0009	1.0408
24	1.0321	1.0363	1.0381	1.0317	1.0227	1.03218	0.00533	0.0107	1.0189
48	1.0194	1.0182	1.0274	1.0154	1.0086	1.0178	0.00609	0.0122	1.0085
72	1.0106	1.0078	1.0112	1.0083	1.008	1.00918	0.00143	0.0029	1.008
96	1.0073	1.0073	1.0091	1.0077	1.0079	1.00786	0.00066	0.0013	1.0076
120	1.0068	1.0073	1.0086	1.0083	1.0078	1.00776	0.00065	0.0013	1.0076
144	1.0072	1.0071	1.0085	1.0077	1.0075	1.0076	0.00050	0.0010	1.0072
168	1.0065	1.007	1.0089	1.0075	1.0076	1.0075	0.00080	0.0016	1.0072
192	1.0062	1.0071	1.0088	1.0071	1.0076	1.00736	0.00085	0.0017	1.007
216	1.0062	1.0076	1.0082	1.0074	1.0075	1.00738	0.00065	0.0013	1.0071
240	1.0066	1.0069	1.0084	1.0070	1.0075	1.00728	0.00063	0.0013	1.0072
264	1.0064	1.0068	1.0083	1.0069	1.0074	1.00716	0.00065	0.0013	1.0072
288	1.0065	1.0066	1.0081	1.0070	1.0078	1.0072	0.00064	0.0013	1.0072
312	1.0062	1.0062	1.0080	1.0067	1.007	1.00682	0.00066	0.0013	1.0065
336	1.0062	1.006	1.0081	1.0068	1.007	1.00682	0.00074	0.0015	1.0068

Figure 24.5. Mean of SG by day chart with the most recent batch added for comparison.

this in a systematic way and eliminate variables to narrow your search. The OE closely reflects the mean (compare data points in fig. 24.5 at $t = 0$ hours), so it is unlikely that too much or too little grain was added. Also, the terminal gravity matches historical data, so it is not likely a mash or boil issue. This brings you to primary fermentation—it took off more quickly than expected. Thinking about the variables that can affect the speed of fermentation (temperature, pitch rate, and DO), you can look back at the batch record and determine if any one of those was out of the ordinary. Just like that, you were able to narrow the search down to a few possible abnormalities in your process. And by checking your records you can narrow it further and use what you find to correct your next batch. As you can see, visualizing how a process performs over time using charts like this is a powerful analysis tool. It allows you to react quickly to abnormalities and prevent them from happening again.

CONTROL CHART BASICS: I-MR

Control charts are based on principles we have already discussed: mean, standard deviation, and normal distribution of data. They are used to confirm measurement systems are valid, spot trends, and determine whether a process is in or out of control. When setting up control charts for routine monitoring, it is critical that the measurement method be in control and producing reliable data. If the method is in control, the data should follow a normal distribution.[3] The analysis of a control chart is a bit more in-depth than the previous chart, but I am confident after we run through this example you can get started on your own and start spotting trends in your data.

There are many different kinds of control chart, each built to handle a specific kind of data set. An **individual moving range** chart (I-mR, also called X-mR) is well-suited for the brewing process because brewers tend to take a single observation (measurement) at a fixed time in the brewing process. For example, a mash pH reading is always taken 15 minutes after mash in ends; in control chart speak, the sample subgroup in this case would be equal to one. Subgroups refer to data points created under the same set of conditions, taken in close succession. In fact, subgroup size is a major factor in selecting which control chart to use (fig. 24.6). We will only cover how to make I-mR charts in this text.

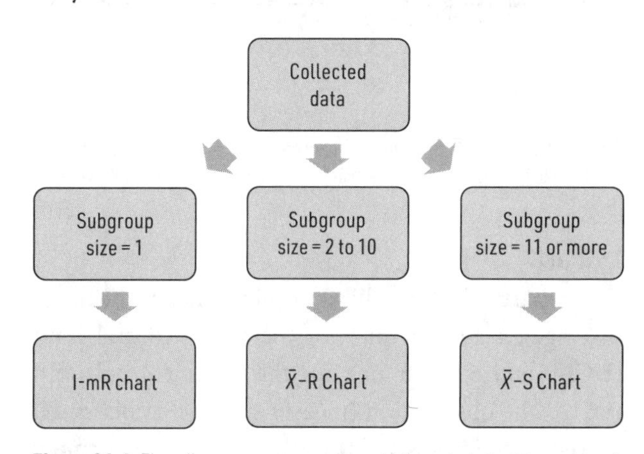

Figure 24.6. Flow diagram recommending which control chart to use based on subgroup size.

An I-mR chart is actually a combination of two charts used in unison to provide a complete picture of a process's behavior. The first is the individuals chart (I-chart). An I-chart displays the individual data points

3 "Control Charting Guidelines for Quality Control," American Society of Brewing Chemists [website], accessed June 1, 2019, http://methods.asbcnet.org/extras/Control_chart.pdf [subscription required].

while monitoring the mean and the shifts in process, if any. The second chart is the moving range chart (mR chart). The mR chart monitors the process variation by taking the absolute value of the difference of the current measurement from the previous measurement. This can be thought of as the precision of the process. Let us take a closer look at each chart and its components (fig. 24.7).

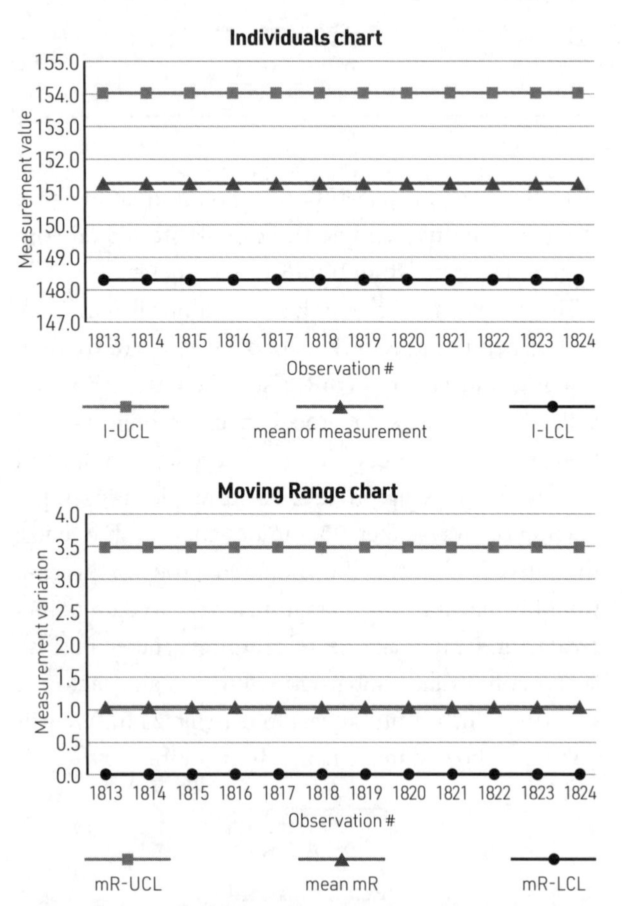

Figure 24.7. These illustrate the components of an I-MR chart. Measurement data has been removed to showcase the mean, upper, and lower control limits.

I-Chart

The I-chart will always have a central line for the mean (average), an upper line for the upper control limit (UCL), and a lower line for the lower control limit (LCL). The observation (measurement) number is displayed on the x-axis and the value of that measurement (x_n) is displayed on the y-axis. The mean for the center line is calculated just as we did in chapter 23, where each measurement is identified by x_n: and n is the total number of measurements

$$\bar{x} = (x_1 + x_2 + \cdots x_n)/n \qquad (24.1)$$

This will give us the value for our central line. Determining the UCL and LCL is a bit more involved. The formulas look like this:

$$\text{UCL} = \bar{x} + E_2 \cdot \overline{mR}$$
$$\text{LCL} = \bar{x} - E_2 \cdot \overline{mR}$$

Let us break down each term in these equations. We already determined $\bar{x}$, which is the mean for our center line. The term E_2 is a control chart constant (see appendix I). Control chart constants give an estimate of the standard deviation of your process based on the size of your subgroup, the variation between measurement, and the kind of control chart you are building. In our case we are building a chart for individuals and our constant E_2 equals 2.660. You will notice in appendix I that the control chart constant E_2 corresponds with a subgroup size of 2. While you are only taking one measurement you are calculating the moving average of two consecutive points, that is, comparing the current state to its previous state.[4]

The natural question to ask here is why are we using an estimate of the standard deviation opposed to the actual calculated standard deviation? Using the estimate of the standard deviation protects us from inflated control limits. If we use the calculated standard deviation we are assuming that data is already in control, which is a circular argument because process control is exactly what we are trying to prove with the control charts. Also, if using the calculated standard deviation, increases in variation will bloat the control limits making it difficult to spot trends.

The next term, $\overline{mR}$, is the mean of the moving range. We can obtain this in two steps. First, take the absolute value of the difference of the current measurement from the previous measurement.

$$mR_n = |x_2 - x_{n-1}| \quad \text{for } n = 2, 3 \ldots n \qquad (24.2)$$

In other words,

$$mR_n = |x_2 - x_1| \quad \text{then} \quad mR_n = |x_3 - x_2|$$

[4] Ramana PV, "What is an I-MR chart?" Six Sigma Study Guide (website), accessed February 21, 2020, https://sixsigmastudyguide.com/i-mr-chart/.

and so on throughout the whole data set. Once we have all of the values for mR_n we can calculate the mean, $\overline{mR}$.

$$\overline{mR} = \frac{(mR_1 + mR_2 + \ldots mR_n)}{n-1} \qquad (24.3)$$

Note the denominator is $n-1$. Because we use two values to calculate mR, n will be one less for calculating $\overline{mR}$ than for $\bar{x}$.

Let us put all of those pieces back together and remind ourselves of the formulae used to calculate the control limit values for an I-chart:

$$\text{UCL} = \bar{x} + E_2 \cdot \overline{mR} \qquad (24.4)$$

$$\text{LCL} = \bar{x} - E_2 \cdot \overline{mR} \qquad (24.5)$$

mR Chart

The second part of our I-mR chart is the moving range chart (mR chart). As stated above, the moving range chart monitors the variation in our data. The variation shows how tightly clustered the data are, which tells us about our process's precision. An mR chart will always have a centerline for the mean of the moving range (i.e., $\overline{mR}$), a UCL, and an LCL. The x-axis value corresponds to the observation number starting at $n = 2$, and the y-axis is the measurement variation (mR_n). Remember there is one less x-axis data point here because mR_n is calculated from two data points.

As you can see, we have already gone through the calculations for some of these mR chart values. The centerline for this chart is equation (24.3) above. This only leaves the UCL and LCL. Referring to the control chart constants in appendix I, we see the constants for the individual moving range are

$$\text{UCL} = D_4 \cdot \overline{mR} \qquad (24.6)$$

$$\text{LCL} = D_3 \cdot \overline{mR} \qquad (24.7)$$

When we take a closer look at appendix I, $D_3 = 0$, which makes sense because the lower control limit represents no variation, thus, the absolute value of a difference between two values where there is no variation will equal zero. The UCL constant, $D_4 = 3.267$, corresponds with a subgroup size of 2, just as we explained when discussing constant E_2 because we are comparing the current state to the previous state.

Building an I-mR Chart Step-by-Step

Using temperature data from the last 12 mashes of a session ale, we will build an I-mR chart to help evaluate if the mash temperature is in control.

1. Organize recorded data, create columns for the following elements: observation #, measurement, mean of measurement ($\bar{x}$), mR, mean mR ($\overline{mR}$), I-UCL, I-LCL, mR-UCL, and mR-LCL.

Your spreadsheet should look something like the following (column headers in row 1 will help us keep track of where the elements fit on our charts):

	A	B	C	D	E	F	G	H	I
1	X-axis	Y-axis (for I-chart)	Center line (for I-Chart	Y-axis for mR-chart	Control limits for I-chart		Control limits for mR chart		
2	observation #	measurement	mean of measurement	mR	Mean mR	I-UCL	I-LCL	mR-UCL	mR-LCL
3	1813	153.2							
4	1814	150.5							
5	1815	152.3							
6	1816	150.7							
7	1817	151.2							
8	1818	151.3							
9	1819	152.6							
10	1820	150.3							
11	1821	150.3							
12	1822	150.4							
13	1823	150.9							
14	1824	151.5							

2. Using the measurement data, calculate $\bar{x}$ (eq. 24.1) [`=AVERAGE($B$3:$B$14)`]. This will make our centerline.
3. Using measurement data starting in cell B4 calculate mR (eq. 24.2) using the absolute value function [`=ABS(B4-B3)`].[5]
4. Using the mR data calculate $\overline{mR}$ (eq. 24.3) [`=AVERAGE($D$4:$D$14)`].
5. Populate the I-UCL and I-LCL columns using the $\bar{x}$ and $\overline{mR}$ data to calculate the control limits using the control limit constant $E_2 = 2.660$ (eq. 24.4 and 24.5). This is [`=$C$3+$E$3*2.660`] for I-UCL and [`=$C$3-$E$3*2.660`] for I-LCL.
6. Populate the mR-UCL and mR-LCL columns using $\overline{mR}$ data and the control limit constants $D_3 = 0$ and $D_4 = 3.267$. Calculate the control limits using eq. 24.6 [`=$E$3*3.267`] for mR-UCL and eq. 24.7 [`=$E$3*0`] for mR-LCL.

[5] If using Google Sheets or Microsoft Excel, this is easily accomplished by using the **ABS** function. Both apps use the syntax =ABS(value).

At this point we have all the data we need to start building the control charts. Your spreadsheet should now look something like this:

	C	D	E	F	G	H	I
1	Center line (for I-Chart	Y-axis for mR-chart	Control limits for I-chart		Control limits for mR chart		
2	mean of measurement	mR	Mean mR	I-UCL	I-LCL	mR-UCL	mR-LCL
3	151.3		1.1	154.0714848	148.4451818	3.4551	0
4	151.3	2.7	1.1	154.0714848	148.4451818	3.4551	0
5	151.3	1.9	1.1	154.0714848	148.4451818	3.4551	0
6	151.3	1.6	1.1	154.0714848	148.4451818	3.4551	0
7	151.3	0.5	1.1	154.0714848	148.4451818	3.4551	0
8	151.3	0.1	1.1	154.0714848	148.4451818	3.4551	0
9	151.3	1.3	1.1	154.0714848	148.4451818	3.4551	0
10	151.3	2.3	1.1	154.0714848	148.4451818	3.4551	0
11	151.3	0.1	1.1	154.0714848	148.4451818	3.4551	0
12	151.3	0.1	1.1	154.0714848	148.4451818	3.4551	0
13	151.3	0.5	1.1	154.0714848	148.4451818	3.4551	0
14	151.3	0.7	1.1	154.0714848	148.4451818	3.4551	0

Now we will start building the I-chart that plots our measurement data.

7. In your spreadsheet insert a chart and select a 2-D line chart with markers.
8. Go to chart tools>design>select data. To the legend entry series (y-axis) add the data for measurement, mean of measurement, I-UCL, and I-LCL.
9. To the category axis (x-axis) labels add the data for observation #.
10. Go to chart tools>design>add chart element and add Chart title, Axis titles, and a legend.

Your I-chart should look like this:

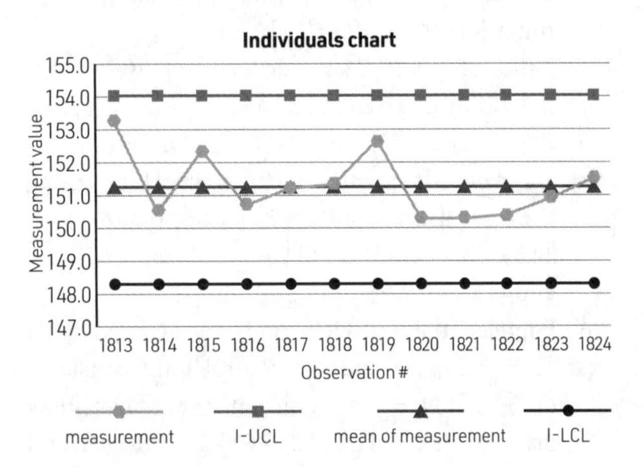

Using similar methods we can plot the mR chart that plots the variation in our measurements

Insert another chart and select a 2-D line chart with markers.

11. Go to chart tools>design>select data. To the legend entry series (y-axis) add the data for mR, mean mR, mR-UCL, and mR-LCL.
12. To the category axis (x-axis) labels add the data for observation #.
13. Go to chart tools>design>add chart element and add Chart title, Axis titles, and a legend.

Putting both charts together your I-mR chart examining mash temperatures on subsequent batches will look like this:

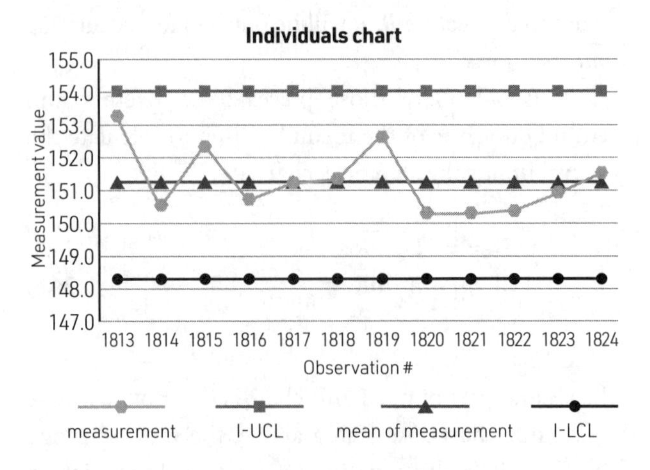

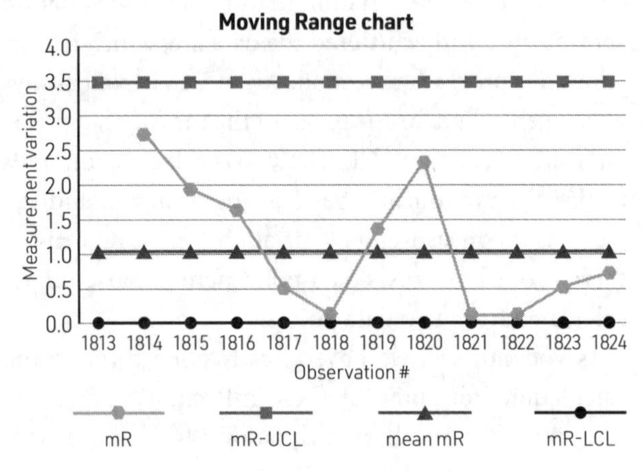

USING CONTROL CHARTS, APPLYING RULES

As we advance, we can compare current data with the "historical data" used to develop this chart. This helps us determine if our process is consistent or out of control. There are several rules that can be applied to control charts that help identify results that are out of statistical control. Four common rules are:

1. Any point outside of ±3 the estimated standard deviation (i.e., UCL and LCL). For a system in control, only 3 out of 1000 points will fall outside of this range due to random error.
2. A run of 8 points all above or below the mean. For a system in control, the probability that this will occur due to random error is $(½)^8 = {}^1/_{256}$.

3. Eight consecutive points running upward or downward.
4. Any non-random patterns (identified by the analyst).

Additional rules can include:

5. Fourteen consecutive points alternating between up and down.
6. Two out of 3 consecutive points above the warning limits (warning limits are ±2 SD).
7. Four out of five consecutive points beyond 1 standard deviation.

By applying these rules to your control charts you can identify if systematic error is present in your measurement or if some part of your process needs attention. In determining how many rules to apply to the chart, more is not always better. As the number of rules that are applied is increased the false alarm rate will increase, which may lead you to conclude that a change has occurred when it has not.[6] No matter what rules you apply the important thing is to be sure and listen to the story that the control chart is telling you about your process.

There are many kinds of control charts available that can be useful to your program. They are broadly like the I-mR example listed above but have been developed to suit particular characteristics of the quality attribute being analyzed. Two broad categories of chart exist, which are based on if the data being monitored is "variable" or "attribute" in nature.Variable control charts are used for monitoring actual process value measurements, like the example we just worked through in "Building an I-mR Chart".

TYPES OF CONTROL CHART

Controlling variables

- I-mR (a.k.a X-mR)
- $\bar{X}$-R
- $\bar{X}$-S

Controlling attributes

- "p" control charts
- "np" control charts

Attribute control charts are utilized when monitoring count data, such as "pass/fail" data, and data that arise as a count in the form of specific numbers (1, 2, 3, 4, . . .)."

p" and "np" control charts: p charts are utilized where there is a pass/fail determination on an attribute inspected, such as true-to-brand sensory data. The p chart will show if the defective proportion within a process changes over the sampling period (the "p" indicates the proportion of successes). In a p chart the sample size can vary over time. An np chart is similar to a p chart; however, with the np chart the sample size needs to stay constant over the sampling period.[7]

Being able to organize and visualize data is one of the most powerful tools in a brewer's toolbox. It allows you to ensure internal standards are being met and processes are staying in control. It also alerts you to a process drifting or changing, allowing you to apply some of the problem-solving techniques we saw back in chapter 22. Being able to get ahead of these changes or failures will save you time, money, and heartache.

[6] "Control Charting Guidelines for Quality Control," American Society of Brewing Chemists [website], accessed June 1, 2019, http://methods.asbcnet.org/extras/Control_chart.pdf [subscription required].

[7] "Types of Control Charts," Presentetioneze.com, accessed May 15, 2019, http://www.presentationeze.com/presentations/statistical-process-control/statistical-process-control-full-details/control-chart/types-control-charts/.

25

CONTINUOUS IMPROVEMENT AND PREVENTIVE MAINTENANCE

The final block on our quality priority pyramid is ready to be placed (p. 15). You have come a long way and put a lot of work into your quality system and its supporting programs. But, you guessed it, your work is not done yet. All brewers should be striving to continuously improve. Even if you are happy with the beer you brewed yesterday, we all have something we can work on. Now that you have the first iteration of your quality system in place, you can look to expand the scope of it and fill in potential holes. Believe me, there is always more to learn.

Continuous improvement is part mindset and part process, and it should complement and strengthen your quality-first ethos. People can get complacent and not want to change, and this new mindset can shake that up and open your eyes to new possibilities and processes. Just because you have done something the same way for years does not mean there is no better way to do it. Being open to suggestion or change is the first step on your road to improvement.

TOOLS AND METHODOLOGIES

There are a plethora of disciplined methodologies you can apply to your continuous improvement goals. Each methodology emphasizes employee involvement and teamwork but may have a unique focus. Choosing a methodology will depend on what you already do well and what your future goals are. Continuous improvement methods worth looking into are:

- **Six Sigma:** A statistical driven approach to reducing defects and costs while increasing customer satisfaction.
- **Lean:** Focuses on workflow to reduce waste, increase process speed, and maximize value.
- **Kaizen:** Aims to improve the business as a whole by standardizing work, improving efficiency, and reducing waste.

At their heart these methodologies are problem-solving tools. Luckily, we already possess a powerful tool that we can apply toward our continuous improvement process. Do you remember the PDSA cycle from chapter 22? You can apply this same problem-solving technique to continuous improvement.

As we saw earlier, the PDSA cycle is an effective way to improve on and dial in your processes. Just like the PDSA cycle, the process of continuous improvement is cyclical because you start with a goal and work toward improving an identifiable metric. Being able to measure your progress in some way is huge, this will prove to yourself that the change is bringing about progress. If adopting a whole methodology is intimidating to you, take a look at some of the problem solving tools employed by these methodologies and use those on a case-by-case basis as you see fit, for example:

- **Kanban:** a visual tool for optimizing workflow
- **The 5 whys:** a thinking tool for identifying root causes

PDSA CYCLE

1. *Plan:* Recognize an opportunity and plan a change.
2. *Do:* Test the change.
3. *Study:* Review the test, analyze the results, and identify what you have learned.
4. *Act:* Act based on what you learned in the check step. If the change did not work, go through the cycle again with a different plan. If you were successful, incorporate what you learned from the test into wider changes. Use what you learned to plan new improvements, beginning the cycle again.

WHERE DO I START?

There is no wrong place to start when thinking of continuous improvement, but I would suggest expanding the scope of your quality system. All of the measurements covered in this book are attainable for you. Start by choosing measurements that fill gaps in your process using the instruments you already have. Once measured and controlled, consider adding a new instrument to monitor further parameters that are important to you.

You can also look to your team for continuous improvement goals. Your team's ideas are free and everyone will enjoy an improved and inclusive work environment. To start, gather your team and take a critical look at your processes from top to bottom. Ask people to identify areas of waste or their pain points and brainstorm solutions as a team. Once the goal is identified, apply one of the problem-solving tools mentioned above and make the change and monitor for improvement. This act will empower your team and make them feel a part of the continuous improvement movement you are spearheading.

Another approach is to look at key performance indicators (KPI) as a starting point. Just because we are craft brewers does not mean we should be inefficient. Begin monitoring KPIs such as:

- brewhouse yield
- overall process yield
- chemical usage
- water and energy usage
- wastewater generated

Identify the low hanging fruit from your KPI list, identify its root cause, and use a problem-solving tool to improve performance.

PREVENTIVE MAINTENANCE

Equally as important as continuous improvement is preventive maintenance. A good quality program will have a comprehensive maintenance schedule for production equipment used throughout the brewing process. If the equipment you use to process your product is inconsistent or unreliable you cannot expect to make a consistent product. Scheduling maintenance can reduce downtime, extend the life of your equipment, and improve consistency across your process. For your equipment, look to the manufacturer's recommendations because each piece will have its own schedule.

CONTINUOUS IMPROVEMENT IN YOUR FUTURE

Continuous improvement is an important part of your quality program. Always be on the lookout for ways to improve your system. The goal is to brew better beer tomorrow than you did today. The general steps are easy: identify an issue and root cause, make a change, monitor, and analyze. As you grow, you may find you need to improve on the few continuous improvement tools I have mentioned here. Look to adopt robust programs like Lean or Six Sigma, which are tried and true methodologies for quality and manufacturing improvement. Not only will you and your team benefit from these improvements, but training in one of these methodologies can also provide job development opportunities.

A FEW FINAL WORDS

You have reached the end of our journey. That was a lot to take in. No doubt you are impatient to begin putting in place a comprehensive quality system for your brewery, if you have not started already. Take a step back and look at your quality system in its entirety. Sparked by the FSMA, you put in place prerequisite programs by writing your GMP and food safety plans, which are the foundation of your quality system. GMP helps ensure that your processing equipment, facilities, and staff are ready to brew a safe and wholesome product. Brewing is just the first part, you want to carry that mentality over to the beer itself, so you developed a food safety plan and identified critical control points that reduce the risk of you releasing adulterated product.

With your foundation built, you started to plan for quality and wrote your first quality manual. This formulated your intentions regarding quality and outlined some rules to help you and your staff reach those quality goals that frame your system. You started building the walls by looking at all kinds of measurements and focusing on the meaning behind them. You identified quality control points and started thinking about their implications, standards, and associated limits.

Finally, you put on the roof by organizing and analyzing the data, setting specifications using your measurements, and looking for ways to keep improving. By the end your house looked more like a pyramid, but I am OK with that. The Pyramids have been standing for centuries and I hope the information presented here will help your brewery stand that long too.

FINAL WORDS OF ADVICE

- Even though the information in this book was presented in a linear fashion it is likely you will have to implement many parts simultaneously. That is OK, just keep your goals clear. Look back at your quality manual and quality statement if you start feeling lost.
- A good scientist is curious, listens, and asks question—you should too.
- Do not solve a problem and skip the root cause analysis. You will likely run into that problem again if you do.
- One measurement is just a number. Repeat your measurements. The first measure is for a number, the second measure is to confirm, and the third measure is for confidence.
- I will paraphrase the best advice ever, no doubt familiar to those of you who started on this journey as homebrewers: "Relax, don't worry, have an independent craft brew."

APPENDIX A

Basic Good Manufacturing Practice Example

Basic Good Manufacturing Practice Example on following page.

Brewery X Good Manufacturing Practice		
Author: Merritt Waldron	**Approved by:** Brewery X head brewer	Version 1.0
Date: 2/17/19		**Effective:** 3/1/19

Purpose

Brewery X is committed to producing beer of the highest quality and safety. To this end, Brewery X has developed, documented, and codified its manufacturing practices to ensure that all beer produced in our facility is wholesome, safe, and delicious.

Scope

This document outlines the framework for Brewery X's GMP and food safety programs. This document works in conjunction with our SOP/SSOPs, production logs, written policies, and training procedures to create a comprehensive food safety and QA/QC management system.

Food safety is the responsibility of each and every employee at Brewery X, and the management empowers every employee to halt production at any time he or she sees a violation of our current GMP or notes a potential food safety concern not currently addressed in this plan.

The practices listed apply to all employees, visitors, and contractors. The rules listed are consistent with Good Manufacturing Practice regulations administered by the Food and Drug Administration (FDA).

GMP Team

The GMP team is the group with responsibility for the development, implementation, and management of the GMP program. The GMP team is comprised as follows:

- Director of Brewing Operations
- Head Brewer
- Director of Quality Control or assigned representative from the QA/QC team
- Plant Engineer or assigned representative from the Maintenance team

Ultimate responsibility for the GMP program lies with the Director of Brewing Operations.

Plant and Grounds

In general, the production plant and grounds are to be constructed and maintained in such a way as to prevent pests, mold or mildew, and contamination of ingredients and beer in process.

Equipment and Utensils

In general, all equipment should be rated for use in a washdown food preparation environment, and all equipment should be easily cleanable and not prevent the cleaning of adjacent surfaces or equipment.

Sanitary Facilities and Controls

In general, incoming and outgoing water must be in specification for its use, mechanicals must be adequate to handle the sanitation loads, and all waste and refuse must be prevented from cross-contaminating ingredients and beer.

Sanitary Operations

In general, all operations at the brewery must ensure effective cleanliness, sanitation, and prevent cross contamination between cleaning and sanitation chemicals and ingredients and beer.

Processes and Controls

In general, all processes are designed to ensure the quality and suitability of ingredients, and to ensure that the beer remains wholesome and safe through manufacture and packaging. Controls are in place to ensure that processes are followed and maintain efficacy.

Personnel

In general, personal hygiene must meet basic standards for cleanliness and good health, and all personnel must follow brewery hygiene guidelines as presented in the employee hand book.

GMP Audits and Compliance Documentation

Past GMP audits are affixed to the end of this document.

All other operation and control documents are recorded "in process" and filed with batch records or in QA/QC lab tracking records.

APPENDIX B

Example Recall Planning and Procedure

GENERAL NOTES

- Be the first to get the story out—that way you control the narrative.
- Acknowledge the issue and take responsibility for it! Do not hide it or try to place blame elsewhere.
- Give all the facts, do not try to hide anything!
- Be open and honest about potential hazards/health risks.
- Communicate a commitment to ensuring public safety.

RECALL PLAN IN EIGHT STEPS

Step 1: Gather details about the recall

- What product is being recalled?
    - Include a photo and batch identification of recalled product in all materials
    - Where can they find the batch # or packaging date?
- What exactly is the issue(s)?
- What are the potential hazards due to this issue?
- When was the product/batch sold?
- Who was the product/batch sold to?
- What should the consumer/retailer do next?
- How are we compensating:
    - customers,
    - distributors,
    - retailers?
- What action are we taking to rectify the situation?
- Who is responsible for each step in this plan?

Step 2: Prepare script

- Draft a press release.
- Be sure to include the information from Step 1 in the press release.
    - It is better to give too much information than not enough.
- All statements should be the same. Deviation from or inconsistencies in the script will lead to confusion and anger.

Step 3: Communicate with government agencies

- TTB
- FDA

Keep detailed records throughout the recall. Be ready to provide the following details when asked:

- Quantity of recalled product
- Details of recall effectiveness
    - What date(s) were consumers were notified?
    - What number of consumers were notified?

    - The number of consumers responding to the recall notice: 75% response rate is unheard of; 50% response rate is very good; 10% response rate means you need more advertising and a better communication strategy.
    - Is the product being returned?
    - Are there regions of distribution that need more attention?

Step 4: Internal communication

- Communicate with all staff.
    - Make sure that all staff members are fully informed and able to communicate about the recall clearly.
- Communicate with distributors.
    - Work with the distributors to locate the recalled product in the market.
        - Who was the recalled product sold to and when?
        - Do they still have the product in stock, or has it been sold to consumers?
            - If sold, what is the retailer's plan for communicating recall to their customers and how can your brewery assist in this communication?

Step 5: External communication

- Communicate with the press.
    - Post a notice to your website that includes a copy of the press release.
    - Send an email to the media in all distribution regions with a copy of the press release both in the body of the email and as an attachment.
    - If it is a major recall, consider:
        - holding a press conference,
        - taking out ads in local press outlets,
        - hiring an outside PR firm.
- Communicate with consumers.
    - Email the entire consumer mailing list.
        - Stick to the script! Use the press release.
    - Post to all social media platforms, for example, Facebook, Twitter, Instagram.
        - Boost the posts and do repeat postings to ensure it is seen.
    - If it is a major recall, consider additional advertising.
        - Poster with information
        - Paid digital ads on news websites
        - Paid printed ads
        - Paid radio/TV ads
        - Snail mail to retail accounts

Step 6: Communicate to outliers

- Reach out directly to those who use your product in their own products.

Step 7: Post-recall analysis

- How successful was the recall?
- How are you ensuring the issue does not occur again?
- Draft a follow-up press release detailing the conclusion of the recall.
    - A recall that is handled well is a marketing opportunity and can build trust and loyalty to a brand.

Step 8: Product re-release

Gather relaunch details:

- What is the timeline for the re-release?
- Where will the product be distributed?
- What changes were made?
- How is the product better than before?
- Why can the consumer trust us?

APPENDIX C

How to Write an SOP

SOP ID: RISING TIDE 001 – HOW TO WRITE AN SOP

Author: Merritt Waldron
Version: 1.0
Approved By:
Effective Date: 06.26.2019

Revision summary **Current - Version 1.0**	**Past - Version (last version number)**
Document created	

Training Documentation

Employee	Trainer	Completion Date	I understand the fundamental principles associated with this task and can perform it effectively as it relates to the condition and safety of myself, others, and the product. (Employee signature)	Approval (Dept. Manager or DBO)

SOP ID: RISING TIDE 001 – HOW TO WRITE AN SOP

Author: Merritt Waldron	
Version: 1.0	
Approved By:	
Effective Date: 06.26.2019	

PURPOSE AND SCOPE
This SOP is written to aid in the development and creation of SOPs. Using this procedure is intended to create uniformity of SOPs between departments, increase the value of SOPs as a resource, and aid in the readability of reference documents.

SAFETY (PPE)	EQUIPMENT	CHEMICAL
Personal protective equipment (PPE) is listed here if required	Supporting tools and parts should be listed here. • Note: the number of tri-clamps and BF valves or things that are inherent to the task that could be covered in training does not belong here	Chemicals needed and their specific concentrations are listed here

SAFETY CONCERNS
Before implementing an SOP, a hazard analysis should be completed and the findings listed here. Learn hazard assessment principles at https://www.brewersassociation.org/educational-publications/hazard-assessment-principles/

REFERENCED DOCUMENTS	TIMING
Reference any other SOPs or tracking documentation relevant or necessary to completing the task Hazard-assessment form	Outline how long should this task should take

RESPONSIBILITIES
List the responsibilities associated with this task and SOP

PROCEDURE
Procedure outlined in full below

SOP RISING TIDE 001: PROCEDURE

WHO WRITES AND UPDATES SOP?

- SOPs can be written and updated by department heads or appointed expert approved by owners.
- All staff members are encouraged to suggest SOP improvements to department heads or the owners. Suggestions will be evaluated with regards to safety, food safety (CCPs), and quality (QCPs), in that order.

PREPARE TO WRITE THE SOP

Map out the higher-level tasks that must be done to complete the task (see example). These high-level tasks will be the bold face type for each part of the task.

EXAMPLE: MAPPING OUT HOW TO MAKE A SANDWICH
• Prep work surface
• Gather ingredients
• Assemble the sandwich
• Plate the sandwich
• Eat the sandwich

WRITE THE FIRST DRAFT OF SOP

- For each high-level task write the steps required for completion.
  - Use concise language
  - **Use bold text preceded by the ✋ icon to draw attention to critical control points (CCPs) or quality control points (QCPs)**

- Use icon to point out when and where information should be documented
- Refrain from adding anecdotal information that can be conveyed during training

EXAMPLE CONTINUED. [MAP OUT HOW TO MAKE A SANDWICH]
Prep work surface
1. Clean cutting board and knife
2. **Sanitize cutting board and knife**
3. Record that equipment was cleaned and sanitized
Gather ingredients etc.
. . .

AUDIT SOP DRAFT

1. With a draft of the SOP in hand, walk through that task step-by-step.
   a. Does the SOP as written accurately represent the order of operations?
   b. Will the task be executed safely and completely as written?
3. Review hazard assessment principles document: https://www.brewersassociation.org/educational-publications/hazard-assessment-principles/
4. With a draft of the SOP in hand, print out and complete a hazard assessment using Hazard-Assessment-Form (file located in Dropbox SOP folder).

WRITE FINAL DRAFT OF SOP

1. With edits and hazard assessment complete, finish the final draft of the SOP.
2. Place the finished SOP into the template. Access the template in Dropbox SOP folder.
3. **Fill in the SOP title page with purpose and scope, safety information, referenced documents, timing, and responsibilities.**
4. With SOP in hand perform a final audit to ensure it works safely and completely.
5. **Get approval signed off by department head or owner.**

TRAIN APPROPRIATE STAFF ON SOP

1. With SOP in hand, the trainer will train staff step-by-step as written.
   a. Explain not only how but why this SOP and its steps are important.
2. **With SOP in hand, the trainer will shadow the trainee until they display complete and total understanding of the task and concepts surrounding it.**
   a. This can be accomplished with a trainer walk-through of the SOP.
3. Once training is complete, trainer and trainee can sign off on training documentation.
   a. Until training is complete and signed off on, the staff member is not allowed to perform the task detailed in the SOP without supervision.
   b. All staff members are responsible for ensuring they are trained on the most up-to-date SOP for all tasks, which is transparently displayed on the affixed training documentation sheet.
   c. SOP updates will be communicated via email and in department meetings.
   d. Management is responsible for prompt training on SOPs and SOP updates.

DOCUMENT CONTROL

- When an SOP is created it is assigned an ID name and number, e.g., "Rising Tide Brewing 001 – How to write an SOP."
- When an SOP is updated the version number must be changed.
  - The old version is archived digitally by department in an archive folder.
  - The new SOP gets a new version number and an updated bulleted synopsis of the update is displayed in the revision summary.
  - The old version is replaced with the new version in the SOP binder.

GLOSSARY OF TERMS
SOP: A step-by-step set of instructions for any repeatable procedure of a complex but routine duty.
CRITICAL CONTROL POINT (CCP): An important point in a production process that can prevent the ingress of potential food safety hazards.
QUALITY CONTROL POINT (QCP): An important point in a production process that can affect the overall quality of the product.

APPENDIX D

QA/QC LAB: Equilibrated Package DO Samples

SOP ID: QA/QC 019 – EQUILIBRATED PACKAGE DO SAMPLES

Author: Merritt Waldron
Version: 1.0
Approved By:
Effective Date: 01.02.2019

Revision summary **Current – Version 1.0**	**Past – Version [last version number]**
Document created	Document created

Training Documentation

Employee	Trainer	Completion date	I understand the fundamental principles associated with this task and can perform it effectively as it relates to the condition and safety of myself, others, and the product. (Employee signature)	Approval (Dept. Manager or DBO)

SOP ID: QC/QA 019 - EQUILIBRATED PACKAGE DO SAMPLES

Author: Merritt Waldron	
Version: 1.0	
Approved By:	
Effective Date: 01.02.2019	

PURPOSE AND SCOPE
Method to equilibrate headspace gases with solution and to read and record DO in preparation for TPO calculation. Equilibrated DO measurements should be taken with the ppb-range (µg/L) DO meter within 15 min. of filling.

SAFETY (PPE)	EQUIPMENT	CHEMICAL
Safety glasses	Scale	N/A
Hearing protection	Timer	

SAFETY CONCERNS
Be aware of pressurized system

REFERENCED DOCUMENTS	TIMING
DO spread sheet.xls	Complete all samples within 15 min. of pulling

RESPONSIBILITIES
Rinse in sample
Record data

PROCEDURE

Canning Run Equilibrated DO Samples

1. Pull can from respective fill head
2. Weigh can (target 494 g ±7 g) and record weight
3. Shake can for 1 minute
4. Place can in piercer and engage
5. Rinse until DO stabilizes and press play to record
6. Record DO (target <75 ppb), CO_2 (target <0.10 drop from tank), and temperature on DO spread sheet.xls

GLOSSARY OF TERMS
DO: dissolved oxygen
TPO: total package oxygen. This is the calculated value that accounts for all DO in package

APPENDIX E

LAB: Forward Pipetting Technique

SOP ID: LAB002 – FORWARD PIPETTING TECHNIQUE

Author: Merritt Waldron
Version: 1.1
Approved By:
Effective Date: 01.02.2019

Revision summary	
Current – Version 1.1	**Past – Version 1.0**
Document created	Document created

Training Documentation

Employee	Trainer	Completion date	I understand the fundamental principles associated with this task and can perform it effectively as it relates to the condition and safety of myself, others, and the product. (Employee signature)	Approval (Dept. Manager or DBO)

SOP ID: LAB002 – FORWARD PIPETTING TECHNIQUE

Author: Merritt Waldron	
Version: 1.1	
Approved By:	
Effective Date: 01.02.2019	

PURPOSE AND SCOPE
This procedure is a forward pipetting technique for use with an air-displacement pipette. Following these steps will increase the likelihood of accurately and repeatedly measuring a specific volume.

SAFETY (PPE)	EQUIPMENT	CHEMICAL
Nitrile gloves Safety glasses Lab coat	Pipette with appropriate tips	Various, pending procedure

SAFETY CONCERNS
Chemicals in use

REFERENCED DOCUMENTS	TIMING
	10–30 seconds per sample

RESPONSIBILITIES
Accurately pipette desired volume for desired application

PROCEDURE

Prepare the pipette tip

1. Set the volume on the pipette
2. Wet the tip 2–3 times
   a. Depress the plunger to the first stop.
   b. Immerse the tip to the correct depth, deep enough to get a full draw but not too deep as capillary action can skew your draw high, and smoothly let the plunger go to its resting position.
   c. Evacuate pipette tip by depressing the plunger.
   d. Repeat steps (a) and (b).
   e. Before drawing you sample completely evacuate tip by depressing to the second stop.

Draw your final sample

3. Holding pipette upright, depress the plunger to the first stop.
4. Immerse the tip to the correct depth, and smoothly let the plunger go to its resting position.
5. Wait about one second for the liquid to flow into the tip.
6. Put the pipette—held between 10° and 45°—against the wall of the receiving vessel and smoothly depress the plunger to the first stop.
7. Wait one second and then depress the plunger to the second stop.
8. Slide the tip up the vessel wall to remove the pipette.
9. Allow the plunger to return to its rest position.

GLOSSARY OF TERMS

APPENDIX F

Common Units and Conversions

	US barrel (bbl.)	US gallon (gal.)	liter (L)	quart (qt.)	cubic foot ($ft.^3$)	cubic inch ($in.^3$)	pound (lb.)	ounce (oz.)	kilogram (kg)	gram (g)
1 bbl.	1.000	31.00	117.3	124.0	4.144	7,161	257.9	4,127	117.0	117,000
1 gal.	0.03226	1.000	3.785	4.000	0.1337	231.0	8.321	133.1	3.774	3,774
1 liter	0.008522	0.2642	1.000	1.057	0.03531	61.02	2.198	35.17	0.9970	997.0
1 qt.	0.008065	0.2500	0.9464	1.000	0.03342	57.75	2.080	33.28	0.9436	943.6
1 $ft.^3$	0.2413	7.481	28.32	29.92	1.000	0.001728	62.24	995.9	28.23	28,230
1 $in.^3$	1.396×10^{-4}	0.004329	0.01639	0.01731	5.787×10^{-4}	1.000	0.03602	0.5763	0.01634	16.34
1 lb.	0.003877	0.1202	0.4549	0.4807	0.01607	27.76	1.000	16.00	0.4536	453.6
1 oz.	2.423×10^{-4}	0.007511	0.02843	0.03005	0.001004	1.735	0.06250	1.000	0.02835	28.35
1 kg	0.008547	0.2650	1.003	1.060	0.03542	61.20	2.205	35.27	1.000	1,000
1 g	8.547×10^{-6}	2.650×10^{-4}	0.001003	0.001060	3.542×10^{-5}	0.06120	0.002205	0.03527	1.000×10^{-4}	1.000

Note: Volume to mass conversions are for water at 77°F (25°C) and 1 atm (101.325 kPa), where the density of water is taken to be 62.2436 $lb./ft^3$ (997.047 kg/m^3). Note that at 68°F (20°C) the density of water is 62.3160 $lb./ft.^3$ (998.207 kg/m^3), so the difference in density between the two commonly accepted definitions of room temperature is negligible.

APPENDIX G

Brewery Lab Start-Up Budget Estimate

With all of the equipment and consumables on this budget you can accomplish a great majority of the methods and measurements outlined in this text. Breweries serving beer to be consumed on the premises only should consider the "All breweries quality instrument list" and the "Consumables" list—for under $4,000 USD you can start to implement your program. Breweries that pack and distribute beer to be consumed outside of the brewery premises should consider the "Packaging brewery instruments list" in addition to the first two lists. This will put you in position to start a robust quality program. The $25,935 USD will be well worth it, helping your brewery grow to 10,000 barrels and beyond if you so choose. This may not be an exhaustive list, and it can be adjusted depending on your personal goals, but is a great starting point.

All breweries quality instrument list	**Price**
Microscope	$290.00
Hemocytometer	$180.00
Hydrometer	$40.00
pH meter (stick type)	$80.00
Volume meter for CO_2 measurement	$1,600.00
Test tubes for HLP tests	$18.00
Pressure cooker (or Instant Pot®)	$80.00
Stir plate	$45.00
Anaerobic chamber	$90.00
Inoculation loop	$10.00
Alcohol lamp	$13.00
Titration burette and stand	$50.00
Beakers	$15.00
Erlenmeyer flasks	$15.00
Media bottles	$15.00
Volumetric flasks	$80.00
Total	**$2,621.00**

Consumables	**Price**
Hsu's *Lactobacillus* & *Pediococcus* Medium – HLP	$106.00
Lin's Cupric Sulfate Medium – LCSM (optional but recommended)	$182.00
Wallerstein Laboratory Differential Medium – WLD (optional but recommended)	$37.00
Lee's Multi-Differential Agar – LMDA (optional but recommended)	$89.00
Disposable transfer pipettes 1mL (300 ct)	$8.00
Methylene blue	$26.00
Serological pipettes 1 mL (100 ct)	$24.00
Serological pipettes 10 mL (100 ct)	$36.00
Gas Pak for anaerobic chamber (20 ct)	$69.00
Petri dishes 60mm (500 ct)	$64.00
Petri dishes 100mm (500 ct)	$52.00
Autoclave sterility indicating tape	$7.00
Whirl-Pak bags	$33.00
Gram stain kit	$45.00
Microscope slides	$10.00
Whatman filter paper	$35.00
Spikes for sensory	$300.00
Nitrile gloves	$10.00
Kimwipes	$6.00
Total	**$1,139.00**

Packaging brewery quality instruments list	**Price**
pH meter (benchtop)	$350.00
Incubator (used)	$350.00
Digital density meter, or densitometer	$2,900.00
Dissolved oxygen meter*	$12,000.00
Orbital shaker table (used)	$300.00
Mini-refrigerator	$100.00
Electrochemical DO meter (ppm range)	$175.00
UV-Vis spectrophotometer	$6,000.00
Total	**$22,175.00**

Notes: Prices estimates are in USD at 2020 prices. Some prices are based on used market value and are indicated by "(used)."

* If budgeting for a combined DO/CO_2 meter plan for $19,000 USD.

APPENDIX H

Control Charts Constants and Formulas

TABLES OF CONSTANTS FOR CONTROL CHARTS

	X bar and R Charts				X bar and s charts			
	Chart for Averages	Chart for Ranges (R)			Chart for Averages	Chart for Standard Deviation (SD)		
	Control Limits Factor	Divisors to Estimate σ_x	Factors for Control Limits		Control Limits Factor	Divisors to Estimate σ_x	Factors for Control Limits	
Subgroup size (n)	A_2	d_2	D_3	D_4	A_3	c_4	B_3	B_4
2	1.880	1.128	-	3.267	2.659	0.7979	-	3.267
3	1.023	1.693	-	2.574	1.954	0.8862	-	2.568
4	0.729	2.059	-	2.282	1.628	0.9213	-	2.266
5	0.577	2.326	-	2.114	1.427	0.9400	-	2.089
6	0.483	2.534	-	2.004	1.287	0.9515	0.030	1.970
7	0.419	2.704	0.076	1.924	1.182	0.9594	0.118	1.882
8	0.373	2.847	0.136	1.864	1.099	0.9650	0.185	1.815
9	0.337	2.970	0.184	1.816	1.032	0.9693	0.239	1.761
10	0.308	3.078	0.223	1.777	0.975	0.9727	0.284	1.716
15	0.223	3.472	0.347	1.653	0.789	0.9823	0.428	1.572
25	0.153	3.931	0.459	1.541	0.606	0.9896	0.565	1.435

TABLES OF CONSTANTS FOR CONTROL CHARTS

	Median Charts				Charts for Individuals			
	Chart for Medians	Chart for Ranges (R)			Chart for Individuals	Chart for Moving Range (R)		
	Control Limits Factor	Divisors to Estimate σ_x	Factors for Control Limits		Control Limits Factor	Divisors to Estimate σ_x	Factors for Control Limits	
Subgroup size	$\bar{\tilde{A}}_2$	d_2	D_3	D_4	E_2	d_2	D_3	D_4
2	1.880	1.128	-	3.267	2.660	1.128	-	3.267
3	1.187	1.693	-	2.574	1.772	1.693	-	2.574
4	0.796	2.059	-	2.282	1.457	2.059	-	2.282
5	0.691	2.326	-	2.114	1.290	2.326	-	2.114
6	0.548	2.534	-	2.004	1.184	2.534	-	2.004
7	0.508	2.704	0.076	1.924	1.109	2.704	0.076	1.924
8	0.433	2.847	0.136	1.864	1.054	2.847	0.136	1.864
9	0.412	2.970	0.184	1.816	1.010	2.970	0.184	1.816
10	0.362	3.078	0.223	1.777	0.975	3.078	0.223	1.777

	Centerline	Control Limits	
Median Charts	$CL_{\tilde{x}} = \bar{\tilde{X}}$	$UCL_{\tilde{x}} = \bar{\tilde{X}} + \bar{\tilde{A}}_2\bar{R}$	$LCL_{\tilde{x}} = \bar{\bar{X}} - \bar{\tilde{A}}_2\bar{R}$
	$CL_R = \bar{R}$	$UCL_R = D_4\bar{R}$	$LCL_R = D_3R$
Charts for Individuals	$CL_x = \bar{X}$	$UCL_x = \bar{X} + E_2\bar{R}$	$LCL_x = \bar{X} - E_2\bar{R}$
	$CL_R = \bar{R}$	$UCL_R = D_4\bar{R}$	$LCL_R = D_3\bar{R}$

TABLES OF CONSTANTS FOR CONTROL CHARTS

	Centerline	Control Limits	
p chart for proportions of units in a category	$CL_p = \overline{p}$	**Samples not necessarily of constant size**	
		$UCL_{p_i} = \overline{p} + 3\frac{\sqrt{\overline{p}(1-\overline{p})}}{\sqrt{n_i}}$	$LCL_{p_i} = \overline{p} - 3\frac{\sqrt{\overline{p}(1-\overline{p})}}{\sqrt{n_i}}$
		If the Sample size is constant (n)	
		$UCL_p = \overline{p} + 3\frac{\sqrt{\overline{p}(1-\overline{p})}}{\sqrt{n}}$	$LCL_p = \overline{p} - 3\frac{\sqrt{\overline{p}(1-\overline{p})}}{\sqrt{n}}$
np chart for number / rate of units in a category	$CL_{np} = \overline{np}$	$UCL_{np} = \overline{np} + 3\sqrt{\overline{np}(1-\overline{p})}$	$LCL_{np} = \overline{np} - 3\sqrt{\overline{np}(1-\overline{p})}$

APPENDIX I
Brew Log Example

Brew Log Example on following page.

Brand			Unique ID #			Brew Date			Fermentor		
IPA			1821			3/14/19			FV 2		
Grain Bill											
lb.	Ingredient					Lot #					Initials
800	Two Row Pale					MBL-2674					MW
220	White Wheat Malt					201-0075					MW
50	Carapils					RO 1201819					MW
Mash targets		Grist (lb.)	Total (gal.)	Strike (gal.)	Flow rate (gal./min.)	Time flow (min.)	Strike temp. (°F)		Target mash temp. (°F)		
		1070	680	315	4.1	15	162		150		
Mash Data		Mash in time	Mash end time	Sach rest end	Base water	Strike (gal.)	Strike temp. (°F)	Mash pH	Mash temp. (°F)		
Vorlauf		Start	End			Lauter	Start	End			
Sparge		Start	End	Target water (gal.)		Actual water (gal.)		Temperature (°F)			
				365							
Runoff		Start	End	Final runnings gravity (SG)		pH	Temp. (°F)	Kettle volume target (gal.)		Actual kettle volume (gal.)	
Boil		Start		Preboil gravity		Preboil pH		Temperature (°F)			
Boil additions											
Time	lb.	Ingredient			AA%	Lot #					Initials
60 min	4	Magnum			15	P90-12256					MW
10 min	5	Cascade			7	P90-30256					MW
whirlpool	5	Centennial			10	P90-69785					MW
	11	Cascade			7	P90-30256					MW
	11	Centennial			10	P90-69785					MW
Whirlpool		Temperature (°F)		Start		End		Volume (gal.)			
Knockout		Start	End	Target temp (°F)	Actual temp (°F)	Oxygenation		Target liters per minute	Target time	Actual liters per minute	Actual time
				68				2	20		
Final measurements		Target OE	Actual OE	pH	Temperature (°F)	FV set temp. (°F)	Target KO volume (bbl.)	Final KO volume (bbl.)			
		1.060					18				

BIBLIOGRAPHY

Acharya, Tankeshwar. "Catalase Test: Principle, Purpose, Procedure and results." *Microbeonline* (blog). October 13, 2013. https://microbeonline.com/catalase-test-principle-uses-procedure-results/.

———. "Streak plate method: Principle, Purpose, Procedure and results." *Microbeonline* (blog). July 16, 2016. https://microbeonline.com/streak-plate-method-principle-purpose-procedure-results/.

———. "Carbohydrate Fermentation Test: Uses, Principle, Procedure and Results." *Microbeonline* (blog). December 10, 2016. https://microbeonline.com/carbohydrate-fermentation-test-uses-principle-procedure-results/.

Alpha Chemical. "Hop Oil Removal Cleaning Procedures." *Alpha Chemical* (blog). November 14, 2018. https://alphachemical.com/blog/hop-oil-removal/.

ASBC. "Beer Methods Beer – 13, Dissolved Carbon Dioxide (International Method)." In *Methods of Analysis*. 8th ed. St. Paul, MN: American Society of Brewing Chemists, 2001.

———. "Beer Methods Beer – 16, End Fermentation (International Method)." In *Methods of Analysis*. 8th ed. St. Paul, MN: American Society of Brewing Chemists, 2001.

———. "Beer Methods Beer – 34, Dissolved Oxygen (International Method)." In *Methods of Analysis*. 8th ed. St. Paul, MN: American Society of Brewing Chemists, 2001.

———. "Beer Methods Beer – 8, Total Acidity (International Method)." In *Methods of Analysis*. 8th ed. St. Paul, MN: American Society of Brewing Chemists, 2001.

———. "Microbiology Yeast – 3, A Dead Yeast Cell Stain (International Method)." In *Methods of Analysis*. 8th ed. St. Paul, MN: American Society of Brewing Chemists, 2001.

———. "Microbiology Yeast – 4, Microscopic Yeast Cell Counting (International Method)." In *Methods of Analysis.* 8th ed. St. Paul, MN: American Society of Brewing Chemists, 2001.

———. "Microbiology Yeast – 6, Yeast Viability by Slide Culture (International Method)." In *Methods of Analysis.* 8th ed. St. Paul, MN: American Society of Brewing Chemists, 2001.

———. "Sensory Analysis Methods Sensory Analysis – 4. Selection and Training of Assessors (International Method)." In *Methods of Analysis.* 8th ed. St. Paul, MN: American Society of Brewing Chemists, 2001.

———. "Wort Methods Wort – 5, Yeast Fermentable Extract (International Method)." In *Methods of Analysis.* 8th ed. St. Paul, MN: American Society of Brewing Chemists, 2001.

American Society of Brewing Chemists (website). "Control Charting Guidelines for Quality Control." Accessed June 1, 2019. http://methods.asbcnet.org/extras/Control_chart.pdf [subscription required].

———. "Lab Safety Checklist." Accessed December 5, 2019. https://www.asbcnet.org/lab/safety/Documents /ASBCLabSafetyChecklist.pdf [subscription required].

Anderson, Haley. "Parts of a Compound Microscope" Microscope Master (website). Accessed January 3, 2019. https://www.microscopemaster.com/parts-of-a-compound-microscope.html.

Aquilla, Tracy. "The Biochemistry of Yeast - Aerobic Fermentation." Brewing Techniques. *More Beer* (blog). July 25, 2013. https://www.morebeer.com/articles/how_yeast_use_oxygen.

Ashurst, Philip R., Robert Hargitt, and Fiona Palmer. "Ingredients in soft drinks." In *Soft Drink and Fruit Juice Problems Solved.* 2nd ed., 29–66. Cambridge, MA: Woodhead Publishing, 2017.

Bamforth, Charles W. "Beer haze." *Journal of the American Society of Brewing Chemists* 57 (1999): 81–90.

———. "pH in Brewing: An Overview." *Technical Quarterly of the Master Brewers Association of the Americas* 38, no. 1 (2001): 1–9.

———. *Practical Guides for Beer Quality: Flavor.* ASBC Handbook Series. St. Paul, MN: American Society of Brewing Chemists, 2014.

Bailey, Ben. "Setting Up a Quality Program." Presentation at the MBAA District New England – Winter Technical, Merrimack, NH, January 20, 2017.

Benedict, Chaz. "Total Package Oxygen 101 – The Difference Between Dissolved Oxygen and TPO." *Tap Into Hach* (blog). June 20, 2012. https://tapintohach.com/2012/06/20/total-package-oxygen-101-the-difference-between -dissolved-oxygen-and-tpo/.

———. "Total Package Oxygen 101 – TPO Calculation." *Tap Into Hach* (blog). June 27, 2012. https://tapintohach.com/2012/06/27/total-package-oxygen-101-tpo-calculation/.

———. "Creating a TPO Validation Standard." *Tap Into Hach* (blog). April 10, 2013. https://tapintohach.com/2013/04/10/creating-a-tpo-validation-standard/.

Bernier, Bob. "I Have Foamy Draught Beer … How Do I Fix It?" *Technical Quarterly of the Master Brewers Association of the Americas* 53, no. 3 (2016): 155–56.

Bible, Chris. "The Principles of pH." Brew Your Own (website). Accessed May 15, 2019. https://byo.com/article/the-principles-of-ph/.

Blake, B.A. and Jenkinson P. "Tests of Crown Seals and Some Practical Applications." *Technical Quarterly of the Master Brewers Association of the Americas* 2, no. 2 (1964): 165–68.

BMG Labtech [website]. "Nephelometer." Accessed May 4, 2019. https://www.bmglabtech.com/nephelometer/.

Bolton, Jason. "Beer and FSMA What You Should Know." Presentation at the Brewery Sanitation Seminar, Falmouth, ME, October 19, 2018.

———. "Prerequisite Programs." Presentation at the Brewery Sanitation Seminar, Falmouth, ME, October 19, 2018.

Braukaiser. "Understanding Attenuation." Braukaiser.com. Last modified 14:48, March 19, 2009. http://braukaiser.com/wiki/index.php/Understanding_Attenuation.

Brennan, John. "Why Are There Buffers in Fermentation?" *Sciencing* (blog). April 25, 2017. https://sciencing.com/there-buffers-fermentation-8377513.html.

Brewers Association. "Brewers' Responsibilities and Obligations under the U.S. Food Safety Modernization Act (FSMA)." Accessed September 15, 2018. https://www.brewersassociation.org/educational-publications/good-manufacturing-practices-for-craft-brewers/.

———. "Good Manufacturing Practices for Craft Brewers." Accessed September 15, 2018. https://www.brewersassociation.org/educational-publications/good-manufacturing-practices-for-craft-brewers/.

———. "The Best Practices Guide to Quality Craft Beer." Accessed October 20, 2018. https://www.brewersassociation.org/educational-publications/best-practices-guide-to-quality-craft-beer/.

———. "Industry Updates, Date Lot Coding" *Brewers Association* (blog). Ferbuary 1, 2019. https://www.brewersassociation.org/brewing-industry-updates/date-lot-coding/.

———. Craft Brewers Guide to Building A Sensory Panel. Accessed June 4, 2019. https://www.brewersassociation.org/educational-publications/craft-brewers-guide-to-building-a-sensory-panel/.

———. Brewery Operations Benchmarking Survey 2018, Accessed June 29, 2019. https://www.brewersassociation.org/statistics-and-data/brewery-operations-benchmarking-survey/.

Brewers Association Safety Subcommittee. *Best Management Practice (BMP) for the Development of Safety Programs in Breweries Volume 1: Hazard Assessment Principles.* Accessed August 3, 2018. https://s3-us-west-2.amazonaws.com/brewersassoc/wp-content/uploads/2018/04/Hazard-Assessment-Principles-BMP.pdf.

Cask Brewing Systems, Inc. "Can Seam Evaluation Training Videos." Web page with three embedded training videos. Accessed on April 1, 2019. https://www.cask.com/service-support/seam-evaluation-videos/.

ChemCollective [website]. "Resource Topic: Acid-Base Chemistry." Accessed February 17, 2019. http://chemcollective.org/activities/topic_page/5.

CMC-KUHNKE, Inc. "Double Seam Troubleshooting Guide." Accessed on April 2, 2019. https://doubleseam.com/.

———. SEAMscan XTS II. Accessed on April 2, 2019, http://www.craftseams.com/seamscan-package-1.

Collins, Cheryl. "Continuous Improvement." *The New Brewer*, July/August 2018.

Cool Cosmos. "Heat vs Temperature." Accessed on December 27, 2018. http://coolcosmos.ipac.caltech.edu/cosmic_classroom/light_lessons/thermal/differ.html.

Cutaia, Anthony J., Anna-Jean Reid, and Alex Speers. "Examination of the Relationships Between Original, Real and Apparent Extracts, and Alcohol in Pilot Plant and Commercially Produced Beers." *Journal of the Institute of Brewing* 115, issue 4 (2009): 318–27.

deLange, A.J. "Understanding pH and It's Application in Small-Scale Brewing." *More Beer* (blog). July 18, 2013. https://www.morebeer.com/articles/understanding_ph_in_brewing.

DoD. *Military Standard Sampling Procedures and Tables for Inspection by Attributes.* MIL-STD-105D. Washington D.C.: Department of Defense, April 29, 1963.

Draft Beer Made Easy. "Carbonation - what you need to know to help you serve draft beer." Accessed on March 12, 2019. http://www.draft-beer-made-easy.com/carbonation.html.

Eldred, Jerry L. "Preparation and Use of an Ice-Point Bath." Metrology 101. *Cal Lab: The international Journal of Metrology* 18, no. 4 (Oct/Nov/Dec 2011): 20–24.

Elert, Glenn. "Temperature." *The Physics Hypertextbook* (blog). Accessed December 27, 2018. https://physics.info/temperature/.

FDA. "Guidance for Industry Questions and Answers Regarding Mandatory Recalls." Rockville, MD: U.S. Food and Drug Administration, May 2015.

———. "Mitigation Strategies to Protect Food Against Intentional Adulteration: What You Need to Know About the FDA Regulation: Guidance for Industry Small Entity Compliance Guide." College Park, MD: Center for Food Safety and Applied Nutrition, Food and Drug Administration, August 2017.

———. "FSMA Final Rule for Mitigation Strategies to Protect Food Against Intentional Adulteration." U.S. Food and Drug Administration (website). Accessed August 20, 2018. https://www.fda.gov/food/food-safety-modernization-act-fsma/fsma-final-rule-mitigation-strategies-protect-food-against-intentional-adulteration.

———. "FSMA Final Rule for Preventive Controls for Human Food Current Good Manufacturing Practice, Hazard Analysis, and Risk-Based Preventive Controls for Human Food." U.S. Food and Drug Administration (website). Accessed August 23, 2018. https://www.fda.gov/food/food-safety-modernization-act-fsma/fsma-final-rule-preventive-controls-human-food.

———. "FSMA Final Rule for Preventive Controls for Human Food - Current Good Manufacturing Practice, Hazard Analysis, and Risk-Based Preventive Controls for Human Food." U.S. Food and Drug Administration (website). Accessed September 20, 2018. https://www.fda.gov/Food/GuidanceRegulation/FSMA/ucm334115.htm.

First Key [website]. "Winning Through Continuous Improvement at Your Craft Brewery." Accessed June 29, 2019. https://www.firstkey.com/winning-continuous-improvement-craft-brewery/.

Folz, Roland. "Non-invasive, selective measurement for CO2 in package expands brewers' quality control toolbox." Poster presented at World Brewing Congress, Denver, CO, August 13–17, 2016.

FSPCA. *Preventative Controls for Human Food*:. 1st ed. Chicago, IL: Food Safety Preventative Controls Alliance, 2016.

Fuentes, Maria. "Hemocytometer Protocol." *Hemocytometer.org* (blog). April 4, 2013. https://www.hemocytometer.org/hemocytometer-protocol/.

———. "Hemocytometer Calculation." *Hemocytometer.org* (blog). April 9, 2013. https://www.hemocytometer.org/hemocytometer-calculation/.

George, Mike, Dave Rowlands, and Bill Kastle. *What is Lean Six Sigma*. New York, NY: McGraw Hill, 2004.

Gerster, Robert A. "Turbidity Control" *Technical Quarterly of the Master Brewers Association of the Americas* 6, no. 4 (1969): 218–20.

Glen, Stephanie. "Empirical Rule: What is it?" *Statistics How To* (blog). November 1, 2013. https://www.statisticshowto.datasciencecentral.com/empirical-rule-2/.

Goineau, Stéphanie. "Les milieux de culture et leur utilisation en contrôle qualité." Slide presentation at Congrès AMBQ –Journée technique MBAA, Montréal, QC, November 21, 2012.

Graphing Tutorial [website]. "Building Bar Graphs." Accessed May 22, 2019. https://nces.ed.gov/nceskids/help/user_guide/graph/bar.asp.

HACCP Mentor. "How to Validate and Verify Your Cleaning Process" [blog post]. https://haccpmentor.com/cleaning/validate-verify-cleaning-process/.

Hach [website]. 2014. "Where to Measure Dissolved Oxygen in the Brewery." Application Note. LIT2148. July 14, 2014. http://www.hach.com.

———. "What is the difference between the turbidity units NTU, FNU, FTU, and FAU? What is a JTU?" Support document TE407. https://support.hach.com/app/answers/answer_view/a_id/1000336.

Hall, Michael L. "Brew by the Numbers – Add Up What's in Your Beer." *Zymurgy* (Summer 1995), 54–61.

Hansberry J. "On-line Beer Bottle/Can Inspection Systems for Quality Assurance and Process Control." *Technical Quarterly of the Master Brewers Association of the Americas* 36, no. 1 (1999): 107–9.

Helmenstein, Anne Marie.. "Independent Variables and Definitions and Examples." *Thought Co* (blog). January 9, 2019. https://www.thoughtco.com/definition-of-independent-variable-605238Isenberg.

Henry D. *Clinical Microbiology Procedures Handbook*. Washington, DC: ASM Press, 2007.

Hershey, Noah. "Brewery Sanitation." Presentation at the Brewery Sanitation Seminar, Falmouth, ME, October 19, 2018.

Holle, Steven R. *A Handbook of Basic Brewing Calculations*. St. Paul, MN: Master Brewers Association of the Americas, 2003.

Kahn, Sal. "Spectrophotometry Introduction," Kahn Academy video 13:06. Accessed April 19, 2019. https://www.khanacademy.org/science/chemistry/chem-kinetics/spectrophotometry-tutorial/v/spectrophotometry-introduction.

Kaiser [pseud.]. "On the Relationship Between Plato and Specific Gravity." *Brewers Friend* (blog). October 31, 2012. https://www.brewersfriend.com/2012/10/31/on-the-relationship-between-plato-and-specific-gravity/.

Katznelson, Revital. "DQM Information Paper 3.1.1: Dissolved Oxygen Measurement Principles and Methods." *The Clean Water Team Guidance Compendium for Watershed Monitoring and Assessment*. Version 2.0. Division of Water Quality, California State Water Resources Control Board (SWRCB), Sacramento, CA, April 27, 2004.

Kessler, Jim, Patti Galvan, and Adam Boyd. "Chapter 3, Density." In *Middle School Chemistry*, 154–243. Washington, DC: American Chemical Society, 2016.

Khouja, Hamed. "Turbidimetry and Nephelometry." King Abdulaziz University (website). Accessed March 5, 2019. https://www.kau.edu.sa/Files/0013791/Subjects/Turbidimetry%20and%20Nephelometry.pdf.

Kiehne, Matthais, Cordt Grönewald, and Frédérique Chevalier. "Detection and Identification of Beer-Spoilage Bacteria Using Real-Time Polymerase Chain Reaction." *Technical Quarterly of the Master Brewers Association of the Americas* 42, no. 3 (2005): 214–18.

Klimovitz, Ray and Karl Ockert. *Beer Packaging*. 2nd ed. St. Paul, MN: Master Brewers Association of the Americas, 2014.

Kotz, John C., Paul M. Treichel, and John R. Townsend. *Chemistry & Chemical Reactivity*. Belmont, CA: Brooks/Cole, 2009.

Krogerous, Kristoffer, and Brian R. Gibson. "125th Anniversary Review: Diacetyl and its control during brewery fermentation." *Journal of the Institute of Brewing* 119 (2013): 86–97.

Kunze, Wolfgang. *Technology Brewing and Malting: 3rd International Edition*. Berlin, Germany: VLB Berlin, 2004.

Leiper, K. A., C.W. Bamforth, and M. Miedl. "Colloidal stability of beer." In *Beer: A quality perspective*, 111–61. Burlington, MA: Academic Press, 2009.

Lewis, Michael and Tom Young. *Brewing*. 2nd ed. New York: Kluwer Academic/Plenum Publishers, 2002.

Li, Hong, Fang Liu, Lidong Kang, and Mingjie Zheng. "Study on the Buffering Capacity of Wort." *Journal of the Institute of Brewing* 122, issue 1 (2016): 138–42.

MacWilliam, I.C. "pH in Malting and Brewing." *Journal of the Institute of Brewing* 81 (1975): 65–70.

Manfreda, John. "Industry Circular Number: 2010-6." Department of the Treasury Alcohol and Tobacco Tax and Trade Bureau, 2010.

Maskell, Dawn L. "Brewing Fundamentals, Part 2: Fundamentals of Yeast Nutrition." *Technical Quarterly of the Master Brewers Association of the Americas* 53, no. 1 (2016): 10–16.

MBAA. "Why Should Breweries be Concerned With Food Safety." Master Brewers Association of the Americas (website). Accessed September 15, 2018. https://www.mbaa.com/brewresources/foodsafety/Pages/WhyFoodSafety.aspx.

———. "Food Safety Modernization Act: Insight for Breweries." Master Brewers Association of the Americas (website). Accessed September 1, 2018. https://www.mbaa.com/brewresources/foodsafety/Pages/FSMA-and-Breweries.aspx.

MBAA Task Force for Food Safety in the Brewing Industry. *Introductory Document for Good Manufacturing Practices (GMPs) as Adopted by the MBAA – Good Brewing Practices (GBPs)*. Accessed August 2, 2018. https://www.mbaa.com/brewresources/foodsafety/Documents/Good%20Brewing%20Practices%20Introduction%202013.pdf.

Math is Fun [website]. "Confidence Intervals." Accessed June 15, 2019. https://www.mathsisfun.com/data/confidence-interval.html.

Math Planet [website]. "Basic information about circles." Accessed March 1, 2019. https://www.mathplanet.com/education/geometry/circles/basic-information-about-circles.

McCabe, John T. *The Practical Brewer, a Manual for the Brewing Industry*. 3rd ed. St. Paul, MN: Master Brewers Association of the Americas, 1999.

McCabe, Kevin. "Is PCR Right for Your Brewery QC Program?" Webinars On Tap. Master Brewers Association of the Americas. Broadcast January 24, 2019. https://www.mbaa.com/education/webinars/Pages/PCR_webinar.aspx.

Microbugs [website]. "Gram Stain." Accessed June 6, 2019. https://www.austincc.edu/microbugz/gram_stain.php.

Milivojevich, Andrew. *Average and Range Charts: Reduce Variation and Save Money*. Accessed July 15, 2019. https://andrewmilivojevich.com.

Moresteam [website]. "X Bar and R Control Charts." Accessed May 16, 2019. https://www.moresteam.com/university/workbook/wb_spcxbarandrintro.pdf.

Murphy and Son Limited [website]. "Last Runnings." Accessed September 16, 2018. https://www.murphyandson.co.uk/resources/technical-articles/last-runnings/.

Niles, Robert. "Standard Deviation." Robert Niles (website). Accessed June 15, 2019. https://www.robertniles.com/stats/stdev.shtml.

Oakland, John S. *Statistical Process Control*. 5th ed. Burlington, MA: Butterworth-Heinemann, 2003.

Ockert, Karl. *Fermentation, Cellaring, and Packaging Operations*. St. Paul, MN: Master Brewers Association of the Americas, 2006.

Oliver, Garrett (ed.). *Oxford Companion to Beer*. Oxford University Press, 2012.

OneVision. "SeamMate System Demo, Double Seam Inspection, Can Seam Inspection Equipment." YouTube video uploaded by OneVision Corporation, July 12, 2017, 3:45. https://www.youtube.com/watch?v=O6j8gyVu5RQ.

Papazian, Charlie. *The Complete Joy of Homebrewing*. 3rd ed. New York, NY: Harper Resource, 2003.

Pellettieri, Mary. *Quality Management: Essential Planning for Breweries*. Boulder, CO: Brewers Publications, 2015.

Pentair. Haffmans CO_2-Selector Non-Invasive CO_2 Measurement. Product Leaflet , 2017.

Phal, Roland. "Beer Turbidity: Reasons, Analytics and Avoidance." Slide presentation, Craft Brewers Conference 2015, Portland, OR.

Pratt, Liz. "Introduction to Sensory Analysis and Beer Flavor." Lecture given at the MBAA Brewing and Malting Science Course, University of Wisconsin, Madison, WI, November 1, 2018.

Presentationeze [website]. "Types of Control Charts." Accessed May 15, 2019. http://www.presentationeze.com/presentations/statistical-process-control/statistical-process-control-full-details/control-chart/types-control-charts/.

Reilly, Christopher. "Detection of Yeast and Bacterial Contamination in Beer: Methods Used in a Craft Brewery." Presentation at the 2016 MBAA Philadelphia Technical meeting.

Rench, Richard J. *Brewery Cleaning: Equipment, Procedures, and Troubleshooting*. St. Paul, MN: Master Brewers Association of the Americas, 2019.

Sammartino, Mark P. "Maintaining Good Yeast Health: A Reflection from Practice." *Technical Quarterly of the Master Brewers Association of the Americas* 53, no.4 (2016): 238–42.

Sheahan, Tim. "Beer Colloidal Stability" *Brewers Journal* (blog). January 24, 2017. https://www.brewersjournal.info/beer-colloidal-stability/.

Slutz, Sandra, and Kenneth L. Hess. "Experimental Design for Advanced Science Projects." Science Buddies (website). Accessed April 19, 2019. https://www.sciencebuddies.org/science-fair-projects/competitions /experimental-design-for-advanced-science-projects.

Souza Dias, Francisco. "Temperature sensors." Coastal Wiki. Last modified June 29, 2019, 21:05. http://www.coastalwiki.org/wiki/Temperature_sensors.

SPX Corporation. "CIP and Sanitation of Process Plant." White paper. Charlotte, NC: SPX Corporation, 2011.

Tague, Nancy R. *The Quality Toolbox*: 2nd ed. Milwaukee, WI: Quality Press, 2005.

Taylor, David G. "The Importance of Control During Brewing." *Technical Quarterly of the Master Brewers Association of the Americas* 27 (1990): 131–36.

Taylor, John R. *An Introduction to Error Analysis: The Study of Uncertainties in Physical Measurements.* 2nd ed. Sausalito, CA: University Science Books, 1981.

Torres, Lauren. "Oxygen Management." Presentation at the MBAA District Michigan Spring Meeting, Comstock Park, MI, 2018.

United States Environmental Protection Agency. *Guidance for Preparing Standard Operating Procedures (SOPs).* Washington DC: Office of Environmental Information, 2007.

Wiesen, Elisabeth, Martina Gastl, Christoph Föhr, and Thomas Becke. "Turbidity and Haze Identification in Beer - An Overview." Poster presented at the 2012 World Brewing Congress. Portland, OR.

White, Chris, and Jamil Zainasheff. *Yeast: The Practical Guide to Beer Fermentation.* Boulder, CO: Brewers Publications, 2012.

White, Chris. "Diacetyl Time Line." Accessed October 14, 2018. https://www.whitelabs.com/sites/default/files /Diacetyl_Time_Line.pdf.

Yourgenome [website]. "What is PCR (polymerase chain reaction)?" Accessed July 25, 2019. https://www.yourgenome.org/facts/what-is-pcr-polymerase-chain-reaction.

Zahm and Nagel Company, Inc. "Part #10005 CO^2 Purity Tester 70-99% in 1.0% and 99-100% in 0.2%." Accessed March 13, 2019. http://www.zahmnagel.com/shop/testing-equipment /part-10005-co%C2%B2-purity-tester-70-99-in-1-0-and-99-100-in-0-2/.

INDEX

Entries in **boldface** refer to photos and illustrations.